INVISIBLE EMPIRE

PRAISE FOR THE BOOK

'Viruses have shaped human civilisation and life on Earth in ways we are only beginning to understand. Pranay Lal's *Invisible Empire* shows us tantalising glimpses of powerful hidden forces that affect each and every one of us. Fascinating and illuminating.'

—SIDDHARTHA MUKHERJEE, Pulitzer Prize-winning author of *The Emperor of All Maladies*

'Most of us think of viruses as agents of disease but they are among the most diverse and abundant organisms in the natural world, right at the boundary between the living and non-living. In this engaging and beautifully illustrated account, Pranay Lal takes us on a grand tour of the world of viruses, revealing their history and the amazing and varied roles they play in nature. Anyone interested in the natural world including young readers will greatly enjoy this book.'

—VENKI RAMAKRISHNAN, winner of the 2009 Nobel Prize in Chemistry

'Viruses are one of life's most powerful and mysterious forces, and shape our world in ways that we don't fully understand. *Invisible Empire* illuminates a world which has so far only been looked at through the narrow lens of disease. This is one of those rare books that can truly change the way you see the world around you.'

—LARRY BRILLIANT, epidemiologist and author

INVISIBLE EMPIRE

THE NATURAL HISTORY *of* VIRUSES

PRANAY LAL

PENGUIN
VIKING
An imprint of Penguin Random House

VIKING

USA | Canada | UK | Ireland | Australia
New Zealand | India | South Africa | China

Viking is part of the Penguin Random House group of companies
whose addresses can be found at global.penguinrandomhouse.com

Published by Penguin Random House India Pvt. Ltd
4th Floor, Capital Tower 1, MG Road,
Gurugram 122 002, Haryana, India

First published in Viking by Penguin Random House India 2021

Copyright © Pranay Lal 2021

10 9 8 7 6 5 4 3 2

The views and opinions expressed in this book are the author's own and the
facts are as reported by him which have been verified to the extent possible,
and the publishers are not in any way liable for the same.

ISBN 9780670095766

Typeset in Plantin MT Pro by Manipal Technologies Limited, Manipal
Printed at Thomson Press India Ltd, New Delhi

www.penguin.co.in

CONTENTS

ACKNOWLEDGEMENTS

It was sometime in January 2020 that the news of a mysterious fever in China began to trickle in. By early February it became a topic of intense discussion with my office chai-mates, Drs Anand Das, Amit Sahu and B.M. Prasad. Very soon it became obvious that our predictions, like everyone else's at the time, were horribly off the mark. As the virus spread across continents, leaving scores of dead in its wake, what also became obvious was the deep animosity towards viruses or really anything microbial, especially among my public health colleagues. While I understood that their hostility was justified, I also knew that the virus–human story is not so black and white. Since my graduate student days, I have understood that microbial communities are like human societies—ecological conditions can keep them at peace or turn them against one another, or against other organisms. However, no amount of my convincing them about how essential microbes as a whole are or how viruses can be beneficial could persuade my otherwise scientific and rational colleagues. Trying to reason with friends and family, too, drew their ire and castigation. How could I be so insensitive, they asked.

In early April 2020, when Namita Gokhale and Sanjoy Roy asked me to give a talk for the Jaipur Literature Festival on the pandemic, I hesitated.

I had very little to add beyond what scientists all over the world were discussing. Instead I offered to speak about the benefits of viruses. Namita and Sanjoy were surprised but agreed. Pratik Kanjilal, my wonderful interlocutor, helped me set the tone and allowed me to approach the topic with sensitivity. A few weeks later, Pavan Srinath asked me to do a podcast for the Bangalore International Centre, and other talks followed. I realised that the story which was so clear to me was not as obvious to others. I went back to my copious notes and pieced together a narrative about the natural history of viruses. This book is a result of that. I have little to offer to those looking for racy accounts of exotic killer viruses emerging from steamy tropical jungles or gory details of past epidemics that changed the course of history. However, those who are curious to know about elements of natural history from the perspective of the microbe-kind, this book is for you. I will feel redeemed if I can at least half convince the reader of the vital contributions that viruses and microbes make to our lives. I have chosen eleven stories from the colourful world of viruses that I found interesting. There are many others that could have easily and rightfully found a place here, and I hope there will be other writers who will find ways to tell those stories in more interesting ways.

This book would not have happened without my wife Vandana Singh-Lal who gave generous advice on every draft. V's comments shaped my early text in more ways than I can tell. As I was struggling to put forth my contrarian views, she helped me find a logical thread for the disparate stories. In these difficult times, she juggled between the various demands of work and family, and she put her writing plans on hold for some time to focus on reading and correcting my umpteen drafts. Thank you, V. *Muito obrigado*, Pradip Krishen, for being such an important part of my life, for being excited about my ideas, for hearing my long-winded stories patiently, and for reading every terrible draft I sent. The book is so much better for the insights you provided. And my children Aria and Avie, angel and cherub, often imps and sometimes devils, for continuing to be patient with, and even excited about, my incessant conversations about viruses while we have been stuck together in unprecedented ways in this rather extraordinary year. Thank you, Avie, for giving insightful comments.

This past year I badgered scientists and experts with telephone calls and requests for online interviews. I sought their advice to see if I was correct in my interpretation as I navigated through a vast maze of scientific literature. I thank Jeremy Barr, Paul Heyman, Tanvi Honap, Nigel Hughes, Rupinder Kaur, Rick McLaughlin, Rhitoban Ray Choudhury, Vinay Sachdeva and Lei Yang for reading my early drafts and providing useful comments. Elliot J. Lefkowitz of the International Committee on Taxonomy of Viruses has generously permitted me to use images from the ICTV database. Sincere thanks to all the illustrators, artists and photographers who have shared their works gratis.

My deep gratitude to Mita Kapoor and the team at Siyahi for supporting my ideas to the hilt; and Meru Gokhale, the consummate author's editor—if only other writers were as lucky! Ahlawat Gunjan did his usual magic with the elegant cover and design. Aparna Kumar stayed vigilant to spot any gaffes that could have crept in while dealing with multiple drafts. Mihir Joglekar, Neeraj Nath at Penguin Random House and my daughter Aria Lal, drew elegant illustrations which have made the book lively.

It is tempting, after having benefited from the help of such generous contributions of thoughts, ideas and editorial advice from experts and friends, to blame the remaining blunders and boo-boos in this book on them. In the interest of honesty, I must resist that temptation. Any mistakes and misinterpretations, that may have inadvertently crept in, are solely mine.

BOUNTY

Let's get this out of the way first: What, *exactly*, is a 'microbe'? Quite simply: any life form that can only be seen under a powerful lens or microscope. And how many microbes exist in the world? If you were to say a trillion to the power of trillion, chances are you will still be well short of the actual number, which is likely to be inestimable, perhaps even unimaginable. Consider this: the ongoing COVID-19 pandemic (short for <u>c</u>orona<u>v</u>irus <u>d</u>isease 20<u>19</u>) started out as a tiny speck in the airway of a wild animal before it passed through hundreds of millions of human lungs, trailing havoc in its path. At the time of writing this, there have been nearly 122 million recorded cases of COVID-19 in the world. A British mathematician has, however, estimated that *all* of the world's circulating SARS-CoV-2 (short for '<u>s</u>evere <u>a</u>cute <u>r</u>espiratory <u>s</u>yndrome <u>c</u>orona<u>v</u>irus <u>2</u>', the causative agent of COVID-19) virus could easily fit inside the confines of a single can of cola!

From what we know, only an infinitesimally small fraction of microbes make us sick, even fewer have the power to kill us. Most simply pass through us, and a few use us as a suitable substrate to make more of their kind. They usually don't bother us in the least and some, in fact, *many*, actually do us good.

In every pinch of undisturbed soil, in every drop of water, there live a billion bacteria and ten times that number of viruses. Every lungful of breath we take contains about a thousand microbes. Most of them are unknown to us. Into this mysterious universe of microbes, come the scientists diving deep, like Captain Nemo, stooped over modern microscopes. They sit gazing at a wondrous world to explore the lives of predators and prey, producers and parasites. Some of the microbes are solitary, a few are in conjugal bliss or are colonisers; many are barely minutes old, the rest are mature, dying or already dead. Each speck of life you see under a microscope has a well-defined role, and in their plurality, they make up habitats no less complex than that of a multi-storeyed tropical forest. The crucial difference is that life in the microbial world plays out several times faster. Microbes come in tantalising forms and often in geometry-defying shapes. They can be translucent and iridescent, twitching or fleeting as one watches them; a plethora of awesome diversity in just a thimbleful of pond water or in soil taken from under leaf litter.

There is no singular world of microbes. Every ecosystem has its own set of curious microbial communities—from the freezing ice of Antarctica to the hot sands of the Sahara, in the placid pond in your neighbourhood to deep-sea volcanic vents. They are everywhere. The surface of a pond teems with numerous photosynthetic organisms. Delve just a foot deeper and the character of organisms changes, while the benthic bottom of the pond and its sediments may have nothing in common with what is on top. A change of seasons or even the time of day can alter the proportions and number of microbes present. Microbial diversity shifts across the surface of the Earth as a consequence of the same three forces that Charles Darwin highlighted to explain the diversity of plants and animals—environment, dispersal and diversification. Microbes mirror the same geographic patterns of diversity as those found in larger organisms, with the variety of microbes (or bacterial species richness) twice as large in the warmer tropics than it is at the frigid Poles.

Microbes come in bewildering shapes, are deeply complex, and in terms of sheer numbers are unmatched by any other life form. Try this for size—it is estimated that there are 100 million times as many bacteria in the oceans (1.3×10^{29}) as there are stars in the known universe. There are more microbes in natural water than in the soil around it, or in the air above. Even on dry land, their numbers are immense and surpass the imagination. The average number of microbes in a single teaspoon of soil (10^9) is as large as the human population of Africa. The human jowl alone, on average, has more than 6 billion microbes that can be made up of more than 600 separate species. A single gram of the stale-smelling yellow grimy film on our teeth, good old plaque, has approximately 10^{11} bacteria, which is about the same number as that of all the humans that have ever lived.

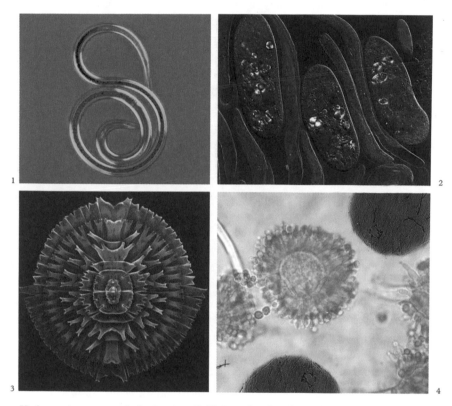

Under a microscope, a pinch of moist soil (1) from the edge of a pond reveals wondrous creatures. A 2-millimetre worm-like nematode wriggles like a giant amidst the soil debris. About 4/5 of all animals on Earth are nematodes—there are 57 billion nematodes for every human on earth. A single gram of soil can house up to several hundreds of these tiny bacteria-feeding worms. If you dive deep using your lens, between grains of soil where there are pockets of water you can see protozoa, like the translucent slipper-shaped Paramecium, *gliding over one another (2). Shift along the slide and you will encounter a solitary cell of a free-living freshwater alga with amazing geometric shapes like this* Micrasterias *(about 0.2 millimetres or 200 microns across [3]). Sharpen your resolution and move further along the plane of the slide and you may encounter a network of fungi with grey heads filled with spores (0.035 mm in diameter) which is* Aspergillus *(4). Alongside it is a bacterial colony of yeast (stained pink), and each of its individual cells are about the size of the knobby head of* Aspergillus. *If you are lucky, you may find a fungal meshwork called* Arthrobotrys *which uses its strands to lasso nematodes which it then slowly suffocates and digests. Each drop of pond water or soil is a complete ecosystem in itself, with its own producers, predators and prey, and parasites, and holds within it the excitement of a tropical safari!*

Communities of microbes are so deeply interdependent that every species relies on one or more of another kind for its survival. The microbial world is organised like a town fair—under the microscope, you will see in a drop of pond water or diluted tropical soil a network of thread-like fungi that run like the tracks of roller coasters and toy trains. Next to them you may see a huddle of amoebae, and close by, congregations of different bacteria each minding their own business, but all in some way cooperating to exploit a particular environment. Pale-green photosynthetic creatures of various shapes and sizes compete for light; some jostle and trade for minerals they seek; a few may fancy their chances to mate or are in the act of gobbling up another while others are in active pursuit; and some still want to be the first to scavenge on the dying and the dead. The effect of this interdependence is deeply significant as it sets the tone of relationships for *all* life on Earth.

The interdependence of microbes with all life that we see around us began in deep time, long before we were even a speck on nature's timeline, as it were. The earliest ancestors of modern microbes kick-started life in a hot and extremely inhospitable world. These minutiae forged partnerships to create colonies and together they broke down and recombined compounds to create new minerals, and especially, free oxygen. For nearly 3 billion years (from 3.5 billion years until 580 million years ago) microbes evolved, and slowly but inexorably and fundamentally, created conditions for multicellular life to emerge. Over these billions of years, microbes developed new cellular processes and chemical pathways which enabled them to utilise minerals and energy, like sunlight and heat, from the earth and water. This is the essence of the balance of the flow of minerals and energy which, at a most basic level, is maintained by microbes. All energy that fuels life rises and ends with microbes. Co-operation and collaboration exist among those with shared interests, whether or not they are related. And there are frequent wars between those who compete for resources.

Among all microbes, viruses are possibly the most enigmatic and certainly the most feared. Although you need special high-powered microscopes to even *see* them, they far outnumber and outweigh all other microbes. So much so that if just the 10^{31} viruses estimated to live in the oceans were laid end to end, they would stretch out to a distance of 100 million light years. Viruses are the simplest of life forms but how do they differ from all other microscopic life forms? Defining a virus with any sort of precision has been a challenge because every time virologists come up with what they consider an acceptable definition, a new discovery comes along to break the rules!

Viruses are mostly just dozens of genes wrapped in a coat or shell of protein. Only very few contain over a hundred genes, and some exceptional ones may have genomes larger than some bacteria do. Virologists believe that there is at least one (if not more) virus that affects every living creature. There are at least 219 viruses capable of infecting us humans. Viruses waft around and show no signs of 'life' until they find a host, and they are labelled 'obligate parasites', which simply means that they do not generate their own energy to reproduce. And the key question is whether or not viruses are 'alive'.

Biologists find it difficult to assign viruses a place within the tree of life. The science of grouping related life forms is called 'taxonomy', and those who practise it are 'taxonomists'. Taxonomists begin with broad, generalised groupings, gradually narrowing their categories down to specifics based on traits of individual species, ultimately to be able to distinguish one life form from another. In modern biological taxonomy, a 'domain' or 'superkingdom' is the highest taxonomic rank of organisms. The current (uneasy) consensus is that all life on Earth as we know it is slotted within three overarching domains called archaea, bacteria, and eukaryotes.

This was not always so. Until as recently as 1977, all life on Earth was divided into a simple binary: eukaryotes (Greek; *eu*: true or good; *karyon*: nut or kernel) that includes more 'sophisticated' cellular life but also contains a few single-celled microbes like the amoebae, or living things with cells that contain membrane-wrapped internal structures like the nucleus—essentially, all living things that are *visible* to us, and a few so small that they can only be seen under a microscope. And prokaryotes (Greek; *pro*: before), which are single-celled organisms that generally lack internal membranes and are *only* visible under a microscope. This binary classification was disrupted in 1977 when evolutionary biologist Carl Woese and his colleagues described a third, more primitive, form of life they labelled 'archaea' which (they said) had been wrongly clubbed with all known minutiae like bacteria. So life, according to Woese, should be divided into three separate barrels or 'domains' instead of two. The first barrel is the eukaryotes, the category of living things to which we belong. The second barrel—the prokaryotes—gave rise to modern bacteria. And the third is the archaea barrel made up of more ancestral organisms that live in extreme environments such as deep-sea vents, metabolizing hydrogen sulphide, methane and organic compounds.

The size, structure and relatedness of these trunks is still debated, although evidence suggests that a common ancestor diverged to give rise to the eukaryotes and archaea. Some taxonomists still lump archaea and bacteria together as prokaryotes. But whether you slot them into two or three groups, viruses impact every single known life form, and yet do not themselves find a place within the tree of life. How so?

Some of us may recall the seven-rank system of classification from our biology class. It was first defined by the Swedish clergyman-turned-botanist, Carl Linnaeus. Although his project of coining a two-word (binomial) Latin name for every species was not entirely original, his contribution in developing clear ground rules of classification was of

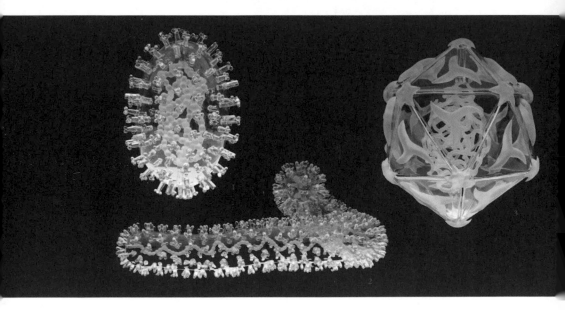

The wavelength of light is often many times larger than the size of most viruses (wavelength of light is between violet (380 nm) to red (750 nm), while most viruses are around 20 nm to 300 nm). These viruses, therefore, do not reflect any light. To 'see' viruses, scientists employ electrons. Electrons are much smaller than the wavelength of light. Scientists bounce off electrons from viruses and capture their image on a monitor to find out their shapes and sizes. The structure and texture of viruses is determined by using x-ray crystallography. Pure, dried and crystallised samples of viruses are exposed to x-rays and a detector records the diffraction (or 'bouncing') of x-rays from the samples. A computer then provides the 3D rendering of the virus by piecing together images of thousands of viral particles taken from different angles. But as no light is involved in this

crucial importance. The microscope had been invented and minutiae were being drawn and discussed, but Linnaeus was unable to differentiate between them and he classified all microbes in the order *Chaos* in 1763 (in modern taxonomy a genus of giant amoebae has been labelled *Chaos*). We still use the Linnaean classification in taxonomy, and according to it, a dog (*Canis familiaris* to taxonomists) is classified using the following steps.

All dogs are grouped into a single genus (*Canis*); several related genera (plural of genus) merge to make a family (*Canidae*); families combine to form an order (*Carnivora*); orders are clubbed together into a class (*Mammalia*); classes make up a phylum (*Chordata*); phyla unite to make a kingdom (*Animalia*); and kingdoms merge to form the topmost classification, the domain. The mnemonic we used when we were learning biology was 'Dear King Philip Came From Granada Spain'. Just say it in reverse to get the initials SGFCPKD!

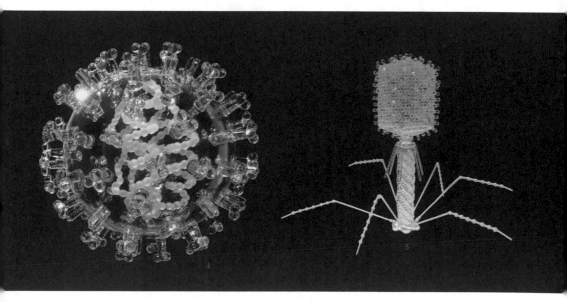

form of 'seeing', there is no colour. For all practical purposes, viruses don't need to have any colour because it would serve no function, neither for mating nor for communicating with one another. Viruses are also not grey. The 'grey' that we see is only the shadow cast by the bombardment of electrons on a monitor. Perhaps only other, very tiny creatures may be able to 'see' viruses, if at all they 'see' each other in their 'light'. The glass models above are perhaps the closest representation of viruses. The avian flu virus (1); the filamentous and curled up Ebola virus (2); the human hand, foot and mouth disease (or enterovirus 71, EV71) which causes mild to severe neurological disorder in children and adults alike (3); the swine flu virus (4); and the spaceship-like T4 bacteriophage that infects and kills the common gut bacteria E. coli (5).

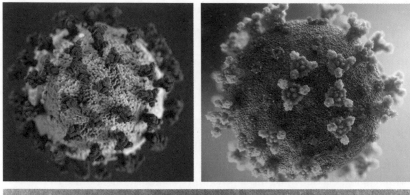

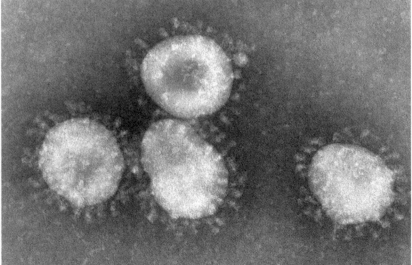

The funky-coloured images of microbes that we get to see in the media are because different colour screens have been used to differentiate viruses from their surroundings and accentuate the surface structures on the virus's protein coat. The colours and contrasts are based on the whims and creativity of technicians and have no real scientific basis. The first popular image of SARS-CoV-2 was created by illustrators at the US Centers for Disease Control and Prevention (CDC) (1). According to The New York Times, the aim of the CDC illustrators was to create 'an identity' for the virus and 'grab the public's attention'. The illustration is a graphic representation of the arrangement of proteins and the design of the virus. The spike (S-) proteins that enable the virus to attach to human cells are shown in red; envelope (E-) proteins that help with adhesion are depicted as yellow crumbs; and membrane (M-) proteins are shown in orange. When the first stories of effective vaccines appeared in the media, a more soothing blue-coloured image was used in print and electronic media articles (2). An electron micrograph (3) shows how we see the virus—in shades of grey. Skilled illustrators then employ visualization software to fill in different colours, textures and lighting effects. In this case, the illustrators chose a stony texture, wanting it to seem like 'something that you could actually touch'. The spikes were made to cast long shadows to 'help display the gravity of the situation and to draw attention'. Although there are more

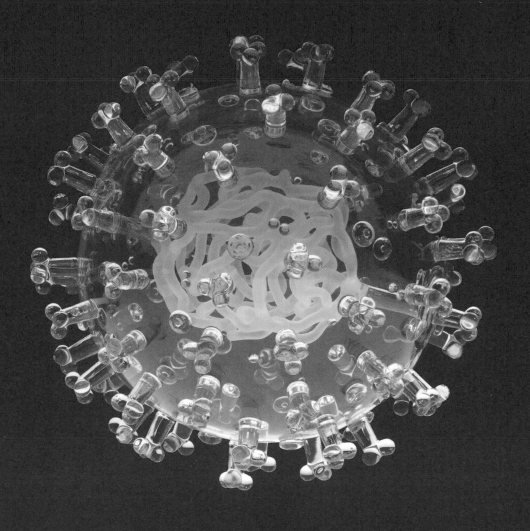

M-proteins than any of the other structures in the virus, they decided to foreground the spiky S-proteins that make the virus 'superspreaders'. Such images aim to communicate health risks but these need to be used carefully to avoid causing unnecessary fear and bias. Perhaps in the not so distant future, technology will enable us to see the true 'colours' of viruses. Until then, an (uncoloured) glass model (above) is perhaps a more accurate representation of what the actual virus is like.

The problem with viruses is that they do not fit into any of the conventionally accepted domains of life—they are neither archaea, eukaryotes nor prokaryotes. Being obligate parasites, viruses stay inert and lifeless until they are able to infect a host cell where they can then steal its energy and start to make more of their own kind. And although they are the simplest life forms (or 'particles' to those who refuse to label them as a form of life), from whatever little we understand about viruses, they are highly diverse and lead complex lifestyles within their hosts.

Apart from their innate diversity and complexity, the way viruses were discovered and viewed also adds to the confusion when classifying them. Discoveries of diseases are made and studied first by pathologists and epidemiologists, who classify diseases on the basis of symptoms whatever the causative agent may be. Take, for example, hepatitis, an inflammation of the liver that is caused by many factors—by downing alcohol beyond moderation, as a side effect of strong medicines, or due to persistent viral infection. And even when the cause of hepatitis is unambiguously diagnosed as a viral infection, the virus affecting the liver also varies in different types of viral hepatitis. Hepatitis A is caused when food or water contaminated with picornavirus infects the liver; hepatitis B is caused by a hepadnavirus (*hepa*: liver; -dna: DNA) that spreads through body fluids just like HIV; hepatitis C is largely spread by sharing contaminated needles or from unsterile tattooing equipment, and is a flavivirus (Latin; *flavi*: yellow) related to dengue. The hepatitis E virus also spreads through contaminated water and food but is distinct as a togavirus (from the Roman dress 'toga', referring to the viral envelope that adorns this particular virus), and many other members of this family are the cause for a number of newly emerged mosquito-borne fevers. To virologists, it makes no sense to lump together all the different viruses that cause hepatitis and give them a common name with different extensions. Yet, that is exactly what has happened, much to their consternation.

Although taxonomists as a tribe are, mercifully, mostly civil to one another, they remain deeply divided on how viruses ought to be classified. Luckily, technology has helped to resolve the confusion for virologists. Using machines that work at a tremendous speed, scientists can now read genomes and compare the genetic make-up across life forms and closely related species. This has helped them to map genomes quickly and to understand the role of individual genes that enable a particular life form to thrive in its specific environment. This systematic study of a large number of viruses, the life forms they infect and their interrelated genomes, is called 'metagenomics'. Metagenomics enables rapid and accurate sequencing of genomes and adds another dimension that enables virologists to understand the evolutionary and genetic relationships between viruses, and also between viruses and their hosts. Based on metagenomics and other advancements, virologists and taxonomists are able to redefine the classification of viruses. Of the millions of species of viruses that may exist in the world, 6590 (up from 5560 in the previous assessment made in 2017) have been named and classified by the apex association that catalogues viruses—the International Committee on the Taxonomy of Viruses (ICTV).

For a long time, virologists and virus taxonomists tried to 'fit' viruses into the conventional seven-rank classification system (Dear King Philip Came From Granada Spain, in reverse). But given the complexity and rapid subspeciation of viruses, this proved to be increasingly problematic. Using new tools of genomics and phylogeny (*phylo*: tree, *gene*: origin), ICTV revised its classification method in May 2020 and expanded it from a seven- to fifteen-rank classification to cover the diversity of the virosphere (note: in real terms there are twelve functional ranks; three ranks—sub-realm, sub-kingdom and subclass—are yet to be populated). In its new avatar, viruses now comprise four realms, nine kingdoms, sixteen phyla, two subphyla, thirty-six classes, fifty-five orders, eight suborders, 168 families, 103 subfamilies, 1421 genera, sixty-eight subgenera and 6590 species. Go ahead and make your own mnemonic!

It is not just viruses that we lack detailed knowledge about—we don't know enough about particularly tiny life forms in general. Part of this problem arises from the fact that much of our knowledge about animal and plant species is dependent upon size. The bigger the life form, the more we know about it. If it is a land animal with a backbone, it most certainly has been carefully mapped and documented. If it is a flowering plant, you can be assured that it is catalogued in an herbarium, potted in a nursery or sports a descriptive sign in some botanical garden or the other. But this is not so for smaller creatures, which, unknown to us, support all the functions of our world. Viruses, especially, are so complex and diverse that they merit being labelled as an empire unto themselves. SARS-CoV-2 has shown us that we know much too little and understand even less about the microbial world, particularly about viruses. If we choose not to appreciate how microbes and viruses influence life around us, we do so at our own peril.

2

A WHOLE
NEW WORLD

Every now and then, discoveries take place that show us our place in
the universe. Leavened by curiosity and sharpened by public debate,
new knowledge and ways of seeing or understanding can act like a
mirror to provide some perspective and remind us that maybe we are
not so special after all. Discoveries over centuries and more recent
revelations have shown us that our universe may not be unique, that
the Sun is just another star, that there are indubitably Earth-like
planets in other galaxies, and, inter alia, that there is an undiscovered
fourth pair of inch-long salivary glands tucked into the space where
the nasal cavity meets the throat! We also realise that our bodies are
not *that* extraordinary when we begin to appreciate and understand
the physiology of other creatures.

1543 was a landmark year. Not because any major catastrophe took place,
or because wars were waged or empires collapsed, but because of two
earth-shattering scientific treatises published that year—*De Revolutionibus
Orbium Coelestium* (On the Revolution of Celestial Bodies) by the Polish
astronomer Nicolaus Copernicus; and *De Humani Corporis Fabrica*
(On the Structure of the Human Body) by the Flemish anatomist and
occasional grave-robber, Andreas Vesalius. Copernicus's work challenged

the belief that our Earth was the centre of the universe and propounded that we revolved around the Sun, not the other way round. Vesalius questioned the old ways of understanding how the body's organs and tissues work, and showed how the interconnectedness of our body parts made us whole. Both these works propelled science towards modernity with great force.

Gradually, findings like these ushered in a new beginning. The Renaissance dawned. The works of Copernicus and Vesalius set off a process of enquiry, of what was within us and what was out there, which would spawn new scientific disciplines over the next century. From this time on, there was a stream of discoveries and inventions that may seem of little consequence today but they made a profound impact on the imagination of the time.

Among them was the use of smooth and rounded glass mirrors. When arranged in a row and placed inside a cylinder, these lenses could bring distant objects closer to become a 'telescope'. Peering through telescopes, astronomers and scientists who followed Copernicus stayed mesmerised by all that they could now see in the night skies. They saw the surface of the Moon in grainy detail, discovered new planets and stars and other celestial bodies, and freely talked and wrote about this. Then, by altering the combination of lenses, tweaking the distance between them and turning them towards objects close at hand, they found they could *magnify* the tiniest specks and littlest creatures. Thus were the earliest microscopes conceived.

The invention of lenses and optical instruments proved to be a major factor in the advancement of science. The early history of the microscope and who its inventors were is not very clear, but a few prototypes were already being sold in Holland and Italy in the 1580s. By the early 1600s, microscopes were being made by instrument makers in England, and like any novelty, they found their way into the homes and libraries of scholars, savants and gentlemen who were curious and scientifically inclined. There were outstanding microscopists from 1650–1750 in Europe who conducted investigations on diverse subjects and made detailed drawings of what they saw. Lectures and demonstrations were organised and observations were exchanged. Public displays were held in homes, at universities and scientific societies, which became quite popular, and laypersons were awed by what they saw—wriggling human sperm, stroboscopic compound eyes of insects, drifting minutiae in a drop of pond water, a bloodsucking flea, and minuscule worms gyrating in wine and vinegar. The invisible world was made visible to everyone. Microscopists demystified many natural phenomena, including our bodies.

Publishers made the most of this growing curiosity and commissioned large books with sometimes even larger illustrations. The Royal Society produced Robert Hooke's lavishly illustrated *Micrographia: or Some Physiological Descriptions of Minute Bodies Made by Magnifying Glasses* in 1665 which became a trendsetter. It was perhaps the most celebrated book of its day. The book revealed an unknown microcosmic world to humanity. Sitting at his desk beside his open window, Hooke became an explorer of a microcosmos that was on us, inside us and around us, and yet it had been completely unknown to us. We can't help but envy the thrill of such endless discoveries made by early microscopists like him. Hooke sliced a very fine section from a piece of cork which allowed light to pass through it, and under the lens, it revealed to him its symmetrical compartments which he termed 'cells' as they reminded him of the honeycomb structure of a beehive. Through a collection of thirty finely grained copperplate illustrations which he published in *Micrographia*, Hooke showed the blow-up of legs and mouths of tiny insects and other common creatures, and of everyday household objects. There were illustrations of a louse, a gnat and a variety of flies, mosquito larvae, a fungus releasing its spores, an ant, silverfish. In fact, Hooke even enlarged commonplace objects like a sharp razor and the point of a needle, both of which appeared decidedly blunt when magnified to that degree! *Micrographia* opened to the reader an uncharted and unimaginable world of microscopic wonders. The book became a conversation-starter in London and spread rapidly across the isles and gradually caught on in scientific circles across the Channel. The captions of the catalogue were just as colourful and incited curiosity in anyone who turned its frond-sized pages—*Of Moss; Of the curious texture of Sea-weeds; Of the stinging points and juice of Nettles, and some other venomous Plants; Of the Seeds of Poppy;* and several other Insects: *Of the Feet of Flyes; Of the Sting of a Bee*; and creatures: *Of the Teeth of a Snail* . . . Even the mundane became marvellous: *Of the Point of a sharp small Needle; Of the Edge of a Razor; Of the fiery Sparks struck from a Flint or Steel; Of Figures observ'd in small Sand . . . and Of Charcoal, or burnt Vegetables.* For the first time, the insignificant took centre stage. With *Micrographia*, Hooke challenged the flawed vision of human creation and contrasted this to the perfection that occurs within every life form in nature—the eyes, mouthparts and bristles on the leg of flies—grotesque, complex and beautiful, all at the same time. The most famous illustration of this catalogue was of

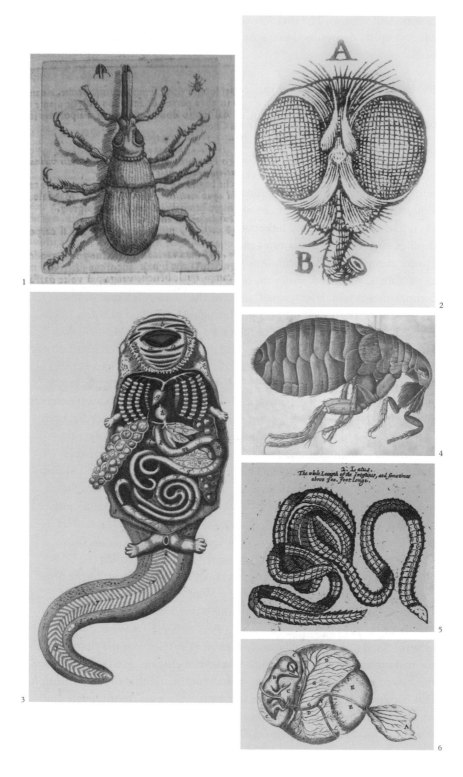

2. Latus.
The whole Length of the Intestines, and sometimes
about 3 or 4 or foot Long.

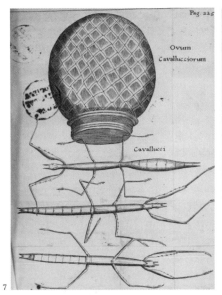

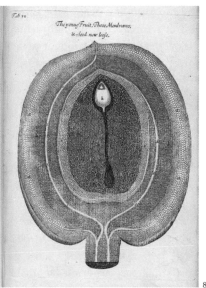

7

8

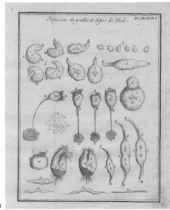

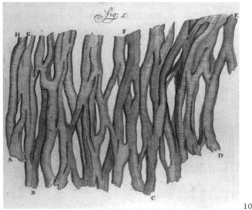

9

10

Among the pioneering microscopists were Francesco Stelluti (1630) who made detailed studies of a weevil (1); Giovanni Battista Hodierna and his magnification of the eye of a housefly (1644) (2); Jan Swammerdam's dissection of a day-old tadpole (1659) (3); Sir Robert Hooke's celebrated flea from Micrographia *(1665) (4); Sir William Ramesey's description of worms that reside within us in his* Helminthologia *(1668) (5); Marcello Malpighi's detail of the development of a chick inside an egg (1672) (6); Francisci Redi's insect egg sac with a young insect (1671) (7); Nehemiah Grew's detailed cross-section of a young fruit (1682) (8); Louis Joblot's study of fungal spores germinating in a heated infusion of hay, which helped debunk the theory of spontaneous generation (1718) (9); and finally, Antonie van Leeuwenhoek's detail of heart muscle fibre (1722) (10).*

the most prolific pest of the time, the human body flea. Like Albrecht Dürer's *Rhinoceros*, the half-a-metre fold-out of the engraving of the flea was reproduced as woodcut prints for centuries to follow. You can still find postcards and bookmarks with the picture of the flea for sale at the British Museum.

Also in England, John Ray's catalogues (1660) of magnified drawings of plant bristles, pollen and insect eyes, mouthparts and legs among other unseen things, and Sir William Ramesey's 1668 charming yet eccentric book on parasitic worms (unappetisingly titled *Helminthologia, or, Some physical considerations of the matter, origination, and several species of worms macerating and direfully cruciating every part of the bodies of mankind; together with their various causes, signs, diagnosticks, prognosticks, the horrid symptoms by them introduced : as also the indications and method of cure, all which is medicinally, philosophically, astrologically, and historically handled*), explained how small parts in bodies made large things work and how microscopic worms were all around and within us. Books and folios like these began to make the Church and the devout royal court feel slighted because they challenged the notion that diseases were acts of God, suggesting instead that they were caused by these insignificant microscopic creatures who were a part of nature.

In Italy, Marcello Malpighi, professor of medicine at the University of Bologna and physician to the Pope, observed and wrote about a wide range of subjects from blood clotting (1666) to how silk was spun by silkworms (1668). His dissection of different stages of development of the chick embryo, its internal organs, the tissues that made them and their different kinds of cells, fascinated scholars across Europe and in England. At the University of Pisa, Francesco Redi's work showed that maggots in wounds arose from minuscule eggs deposited by flies (1668). These were among the first efforts to challenge the notion that the descent of life could take place in the absence of similar organisms (the theory of 'spontaneous generation').

In Holland, monumental discoveries were made by Dutch microscopists. Antonie van Leeuwenhoek was perhaps the greatest microscopist of this time. He started as an apprentice in a cloth merchant's store where magnifying glasses were used to count the threads in silk, linen and cotton, and weaves in jacquards and carpets brought from distant lands. The finer the weave and the higher the number of knots, the greater the value of the cloth. When he started his own drapery establishment, van Leeuwenhoek had to haggle with merchants on the price of these fine fabrics and he devised his own methods of counting warps, wefts and knots which became the standard for the textile industry. Van Leeuwenhoek taught himself new methods of grinding and polishing tiny lenses of great curvature which provided higher magnification, and built his own hand lenses and microscopes, the finest known at the time. It is believed that in his lifetime, van Leeuwenhoek made over 500 microscopes to view specific objects in nature.

In 1668, van Leeuwenhoek made his only trip to London, where he probably saw a copy of Hooke's *Micrographia*. The book aroused his interest in the use of microscopes to discover nature and objects around him. In 1673, he reported his first observations—of a bee's mouthparts and sting, a louse and a fungus—to the Royal Society of London. His correspondence on the existence of single-celled animals—most likely, protozoans in pond water—was met with scepticism at first but was accepted later when other members of the Society also peered through microscopes and confirmed his findings. Van Leeuwenhoek made intricate drawings of cross-sections of cells in leaves and stems, observed the circulation of red blood cells in capillaries, uric acid crystals that caused gout and banded patterns in muscles. His incessant letters about his discoveries enchanted readers of the Royal Society, and in 1680, he was elected a member.

Van Leeuwenhoek was fascinated by infinitesimally tiny organisms in unimaginable numbers moving about very prettily and completely below the threshold of unaided perception. He found corkscrew-shaped microbes that snaked within his dental plaque and, exploring other parts of his body, discovered multitudes of other tiny living things. Van Leeuwenhoek estimated that more creatures were living inside his mouth than there were people living in all of the Netherlands—an idea that disgusted his friends, but by now he was passionately curious to find out more about his body and nature. He was among the first people to understand that we carry within our bodies a menagerie of our own. In every animal, plant and even inside some microbes, he realised there lives a vast community of teeming multitudes. By the time he died in 1723, van Leeuwenhoek had written 192 letters to the Royal Society, all of which were published. His most important discovery was how sperm penetrated the egg and caused fertilization, which generated an interesting and sometimes quite intense debate on both sides of the Channel.

Another great Dutch microscopist was Jan Swammerdam who showed the contraction of human muscle (1658) and its functioning under the

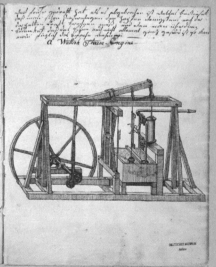

3

The Industrial Age propelled England far ahead of the rest of Europe and there was anxiety in the German states about their ability to match England's manufacturing prowess. In the spring of 1791, a group of businessmen in Mannheim, Germany, decided they needed steam engines to power their grinding mills and to excavate coal and lignite. The technology for the engine lay in England. The German businessmen then sponsored a covert mission to England and recruited a spy for this purpose—a bright young man named Johann Georg Reichenbach, son of an artillery lieutenant and drill master, who had a remarkable aptitude for mathematics and anything mechanical, with draughtsmanship skills to match. Reichenbach's mission was to get hold of the design of the steam engine from Matthew Boulton & James Watt manufactory in Soho and Birmingham (1). Reichenbach became an apprentice in this factory and diligently reproduced the design of these machines in his diaries (2) with the intent to assemble a fully functional steam engine upon his return to Germany. Although he succeeded in getting these designs, there was a falling out among the businessmen in Mannheim and the idea of making the first steam engine outside of England faded away. Reichenbach's escapade is considered by economic historians to be the first case of industrial espionage. On a subsequent visit to England, Reichenbach visited workshops and shops that sold marine and optical instruments, where, once again, he memorised the designs and shapes and took copious notes. He befriended craftsmen and stole ideas and designs from leading theodolite makers. When he returned from his final visit, Reichenbach began designing optical instruments, especially precision-surveying equipment. One of his first ventures was to make the cumbersome English theodolite, used in military expeditions and town planning, more compact and accurate. He invested his energies in improving the quality of glassmaking in Bavaria, and by the late 1780s, Germany was making some of the world's best microscopes. In 1817, Reichenbach built a 25-kilometre-long pipeline with a steam-driven pump to transport brine. Germany was now well on its way to becoming a chemical superpower. For his contributions to the optical sciences, Ludwig I commemorated Reichenbach (left), alongside his friend and pioneering microscope-maker Joseph von Fraunhofer (right), in a one thaler coin minted in 1826 (3).

lens. The French mathematician Louis Joblot demonstrated in 1718 that a heated infusion of farmyard hay quickly grows minute creatures that 'developed from eggs dispersed in the air', which once again went against the commonplace belief that life emerged in a spontaneous way.

With each improvement in the resolution of lenses and the design of microscopes, physiologists and naturalists began to see deeper and farther, and helped clear some of the obfuscation associated with superstition and blind faith. The microscope became a weapon for scientific validation and teased the monolith of the natural sciences into separate strands of varied disciplines. Some early innovative microscopists became pioneers of new branches of knowledge. Redi, for example, became known as the founder of experimental biology and the father of modern parasitology; Swammerdam was designated the father of entomology; Malpighi the founder of modern anatomy and histology; and van Leeuwenhoek the father of microscopy.

About 150 years after van Leeuwenhoek saw bacteria on skin and in dental plaque, a German naturalist named Christian Gottfried Ehrenberg labelled it as that (Greek; *baktéria*: little stick) in 1838. Two decades later, the great French scientist Louis Pasteur became the first to isolate and grow bacteria, and established the brand new field of bacteriology, bristling with potentially new ways of understanding microbes.

The trade in microscopes was largely craft-based—it depended on the quality of glass, mechanical techniques and lens-making skills, all of which were needed to make the best microscopes. After early innovations in Italy and Holland, the best optical instruments were being made in England which remained, for the next 150 years or so until the mid-eighteenth century, the centre of the optics industry. However, incremental advancements began to take place in Germany which then became a hub of fine microscope-makers. By 1811, leading scientists like the English astronomers Frederick William Herschel, John Goodricke and Thomas Hutchins, as well as the best microscopists in Europe and America, had gravitated to using German lenses and microscopes.

As microscopes improved in resolution and acuity, they ushered in a race to identify microbes that caused human disease. Some pathologists began to suspect that there were infective agents even smaller than bacteria that could not be seen under an ordinary microscope. These tiny organisms could neither be isolated nor cultivated, although it was clear that the fluid in which they lived could infect another. In 1857, an agricultural

1

2

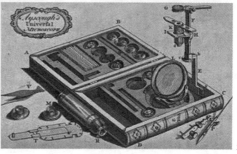

3

In the early sixteenth century, hand lenses were used to magnify small creatures, as depicted in this woodcut which shows a man studying a butterfly (1). Gradually, by the mid-sixteenth and early seventeenth century, microscopes were being used to study body parts and manifestations of disease (2). The image shows an oversized Campani's microscope on the left, perhaps to advertise it, and on the right is a surgeon using the microscope. By mid-seventeenth century, instrument makers in England were making tripod-style microscopes from wood, pasteboard and leather, while Italian designers made smaller and ornate microscopes from turned wood, ivory and brass. Gazettes and handbills advertising the sale of assemble-your-own-kits began to appear in England and Germany in the 1830s (3). In England, the craze for microscopes led to the formation of a society of amateurs and scientists called the Microscopical Society of London in 1839. In 1866, it gained its royal charter and was christened the Royal Microscopical Society.

Microscopes became so popular that in Europe there were public displays like in this street scene from Amsterdam where a travelling showman with his Delabarre brass microscope is showing animalcules in a drop of water to children (oil painting by an unknown Dutch painter, ca. 1840). In 1878, Charles-Emmanuel Sédillot, a French military physician and surgeon, used an improved version of the Delabarre microscope to examine the multitude of organisms that infected wounds and labelled them 'microbes' (Greek; mikro-bios: *small or short-lived*).

In times when there were no epidemics, the public enjoyed watching microbes that lived around them, on them or even within them. The coloured etching (above, titled 'A Monster Soup', commonly called Thames Water, by William Health, 1828) shows a fashionably dressed woman dropping her porcelain cup in disgust after seeing monsters swimming in a drop of Thames water through a microscope. In the 1820s much of the drinking water for Londoners came from the Thames, and the sewers emptied into the river. Between the date of this caricature (1828) and the completion of the sewerage work in the 1860s, London suffered two major cholera epidemics (1832, part of the global cholera pandemic, and a more local one in 1854). Looking at a drop of water though a microscope was a popular entertainment provided by travelling showmen who carried the microscopes around in cases on their backs, but it was caricatures like this one which reminded citizens of the link between dirt, disease and death.

student discovered an unknown disease on a tobacco farm and reported this to Adolf Mayer, director of the Wageningen Agricultural experiment station in the centre of Holland. Mayer studied it for several years and ruled out the possibility of it being a nutritional disease. Not able to find a bacterium, fungus or nematode in the plants, Mayer took an extract from the stems and leaves of diseased plants, mixed it with water and found that when spread over healthy tobacco plants, it caused an infection. When passed through a filter paper, the fluid remained infective, but when passed through a double filter paper, it no longer caused the disease. Heating the infective fluid to 80 °C for an hour also killed its infectiveness. He concluded it was very likely a disease caused by bacteria that were far too tiny to be visible through any means then known.

The disease and its agent remained unclassified until 1890 when Dmitri Ivanovsky of St Petersburg University was sent out by the Department of

Agriculture to study a disease afflicting tobacco plantations in Ukraine and Crimea. Like Mayer before him, Ivanovsky found the sap to be infective. He passed diluted sap through a specialised filtering device called 'Chamberland filter candles' and it remained potent to infect healthy plants. At first, he thought that the filters were defective, but eventually, like Mayer, Ivanovsky, too, concluded that the infective agent was a very tiny bacterium. It is Ivanovsky who is credited for discovering the first virus, not Mayer.

The term 'virus' was actually coined by a third scientist who was studying the cause of mottling in tobacco leaves. Martinus Willem Beijerinck, son of a tobacco dealer who went bankrupt, did his doctoral dissertation on plant 'galls' at the University of Leiden in 1877. He began teaching at the Agricultural School in Wageningen where he met Mayer, and they exchanged ideas on Mayer's early observation and experiments in 1885 on the transmission of tobacco mosaic disease. Like Mayer, Beijerinck, too, was unable to narrow down the data and discover the pathogen and he abandoned this pursuit and turned his attention to other things. However, in 1895, Beijerinck joined Delft Polytechnic as professor and resumed his study on diseases of the tobacco leaf. Just as Ivanovsky had done before him, Beijerinck reasoned that the cause of tobacco mosaic disease might be a bacterial toxin (similar to the one produced by tetanus bacilli, specifically an exotoxin [exo: outside]) where the bacteria, although present in the body, is *not* located at the site of the symptoms. When Beijerinck experimented using a very small amount of it in both wet and dry form, he discovered it could infect numerous tobacco leaves. Exotoxins are proteins and easily rupture when dried or heated, so he concluded that this could not be of bacterial origin. Beijerinck's tobacco extract survived in cold winters and warm summers in the soil, and on dried sun-cured tobacco leaves, and remained infective. Despite repeated efforts, Beijerinck could not find and identify the causative agent and he concluded that the disease was not caused by bacteria but by a *contagium vivum fluidum* (Latin; living fluid infection), a clunky term that he dropped when he adopted the archaic Latin term 'virus'.

To the etymologically inclined, the word 'virus' is in some sense the opposite of itself. The root of the word derives from Sanskrit (विष, *vish*) and in its Latinised meaning, denotes venom or poison. In Middle English, the use of the word got expanded and 'virus' came to mean semen. The word, therefore, signifies both birth and death at once. In a modern context too, 'going viral' suggests that a message or image quickly reaches untold numbers of people, much in the manner in which an outbreak transmogrifies into a pandemic.

Some time before the word 'virus' was coined, a major discovery had taken place in France. On 4 July 1885, a nine-year-old boy named Joseph Meister from Meissengott in the region of Alsace was savagely bitten by a rabid dog fourteen times in all, on his hand, legs and thighs. The local physician had heard about Louis Pasteur's promising work with rabies in dogs and urged the Meisters to take their son quickly to Paris. Mrs Meister and Joseph reached Pasteur's clinic in Paris in the afternoon of 6 July. Pasteur consulted two leading physicians and friends, Alfred Vulpian and Jacques-Joseph Grancher, and considered vaccinating the boy with an early vaccine he had been testing on dogs, for which the results had been promising. Meister's mother was uncertain but Grancher convinced her that the boy was doomed unless Pasteur tried the untested vaccine, and Meister's mother begged Pasteur to save her son. A close collaborator, friend, a famous bacteriologist in his own right and co-discoverer of the cure for diphtheria with Pasteur, Emile Roux chose not to participate in this trial, fearing adverse consequences. He was censorious of the haste with which Pasteur, Grancher and Vulpian were proceeding. He believed that all three of them were disregarding more humane considerations in their unswerving pursuit of scientific knowledge, and perhaps, commercial gain, for the institute. Pasteur had been working on a vaccine to protect dogs against rabies and had shown that it worked successfully in protecting fifty dogs on whom he had tried his vaccine. But the results had not yet reached the larger scientific community or the public. Pasteur's initial aim was to inoculate dogs and not humans with the vaccine until Meister and his mother arrived unexpectedly at his clinic. Pasteur knew that, like polio, rabies infects the spinal cord. He had taken spinal cords from people who had died from the bites of rabid dogs and had serially passaged the rabies 'agent' into dogs and then into rabbits. He had extracted a rabbit's spinal cord and dried it in a flask to protect it from moisture and bacterial contamination. He then crushed the dried spinal cord into a powder and reinfected the spine of a fresh rabbit, repeating the process several times until the symptoms of the disease had become completely benign. Vaccine designers call this process 'attenuation', a gradual reduction in the severity of diseases in a living virus. Today's vaccines for measles, mumps and rubella (MMR) are all examples of similar live, 'attenuated' virus vaccines.

Pasteur's 'vaccine' saved Meister's life. Remember, this was a time before viruses had been isolated or even conclusively identified. What this means is that effective vaccines against a viral disease like rabies were made even before the agent causing the disease had been found!

At 8 p.m. on 6 July 1885, Master Joseph Meister, a young shepherd from Alsace was inoculated by a rabies vaccine prototype. A commemorative frieze at the Pasteur Institute shows Dr Jacques-Joseph Grancher (seated in the image, above) talking to a girl and her mother. As a chemist, Louis Pasteur (arms akimbo to the left of Dr Grancher) was not authorised to perform medical procedures. The vaccine was made using the spinal cord from rabid rabbits that had been air-dried for increasingly shorter periods (15, 14, 12, 11, 9, 8, 7, 6, 5, 4, 3, 2, 1 day), the record of which is seen in Pasteur's own handwriting (opposite page). Before this, no one who had developed symptoms of rabies had survived. After thirteen days of serial low to high dose vaccinations, Meister's symptoms disappeared and he recovered.

Although Pasteur had found a cure for rabies, he had not looked too closely for the pathogen or, for that matter, even checked if the rabies-causing pathogen could pass through Chamberland's fine filters. Ironically, the Chamberland filter, also known as the Pasteur-Chamberland filter, a foot-long metal tube packed with crushed porcelain, was developed by Pasteur's assistant, Charles Chamberland, in 1884. Had Pasteur or Chamberland run the rabies-causing pathogen through their porcelain filter, one of them would almost certainly have isolated the rabies agent and become discoverers of the pathogen ahead of Ivanovsky, or would have had the pleasure of giving the agent a name before Beijerinck did.

In 1903, Ivanovsky published a paper describing 'abnormal crystal intracellular inclusions' in the host cells of affected tobacco plants that he proved were capable of reinfection. By this time, even though they still could not be seen, more and more new viral agents were being isolated. In 1898, the same year Beijerinck's work was published, foot-and-mouth disease in cattle became the first animal illness linked to a filterable agent. In 1901, American army doctors studying yellow fever in Cuba also found a mosquito-vectored pathogen that was filterable. In 1935, American

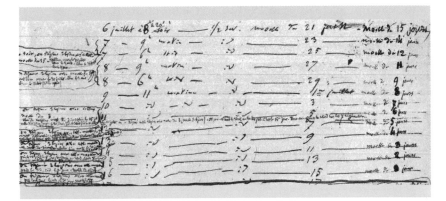

The results of the trial and a second case were so dramatic that by October 1885, members of the French Academy of Sciences advocated that it was 'necessary to organise this treatment for everyone' and 'this is a memorable day in the history of medicine'. Philanthropic contributions poured in, and by 1888, the Pasteur Institute was founded. By then, about 1200 patients had been vaccinated and mortality from rabies declined to just 1 per cent in those who received the vaccine after being bitten. Having survived rabies, a grateful Meister devoted his life to Pasteur's laboratory and worked there as a janitor. When Pasteur died in September 1895, he was buried in a crypt at the institute, and Joseph Meister remained its faithful custodian. It is said that during the German occupation of Paris in 1940, Meister chose to take his own life rather than surrender the keys to Pasteur's memorial, the man who had saved his life and transformed medicine forever.

biochemist Wendell Meredith Stanley was able to crystallise the tobacco mosaic virus (TMV) and show that it remained infective. This property helped in isolating and concentrating the virus.

Meanwhile, microscopes were becoming increasingly more powerful thanks to the application of physics, mathematics and electronics. Physicists discovered that beams of electrons behave as waves, with wavelengths shorter than visible light. This opened up new opportunities to see the unseen in, literally, a 'different light'. By the early 1930s, scientists discovered that when a steady stream of electrons are focused on an object using a cathode ray tube or cathode gun, it creates an image of extremely small objects by deflecting electron beams which can be captured on a screen. They also discovered that a magnetic coil could be used as a 'lens' to focus electron beams, and thus, began the development of the electron microscope.

The first people to reveal they had invented a working electron microscope were German physicists called the Berlin Group. The Group was led by Ernst Ruska, a young doctoral student, and his mentor, Max Knoll, at the Technical University of Berlin, who presented the prototype

of the electron microscope in 1931 as part of his doctoral thesis. The
Berlin Group of scientists worked collaboratively with Siemens (with
limited support from Zeiss, a leading optical instrument company)
to apply for a patent and make it marketable. The Group tested its
equipment by gazing at household objects like the tip of a fountain pen.
They began to publish photographs of how electrons cast their shadows
on a monitor. Among the earliest photographs they published was the
lacy wing of a housefly in 1935. Two years later, they recalibrated their
equipment and could now 'see' even smaller things like diatoms, bacilli,
even a single iron filing.

The first 'proper' electron microscope was launched in 1938. And in
1939, a crystal of TMV was placed under the electron microscope and
scientists saw for the first time slinky rods arranged neatly in bundles, or
'cigarette-shaped particles', as the Berlin Group reported it. The tobacco
mosaic virus, thus, became the first virus to be 'seen' by us.

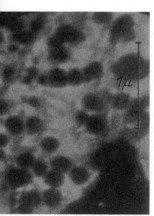
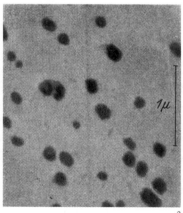
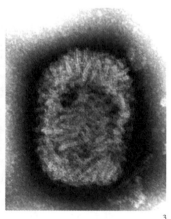

1   2   3

The first virus to be 'seen': *In 1931, Max Knoll and his students, brothers Ernst and
Helmut Ruska, developed an early prototype of the electron microscope which could magnify
objects only seventeen times (17x), and was not much better than a simple microscope.
However, within months after constant tinkering and recalibration, they were able to
increase its magnification to 400x. This rapid development was in stark contrast to the
sluggish evolution seen in the light microscope. In 1933, Ernst Ruska, persevering in the face
of difficulties, built a working prototype and a precursor of the modern electron microscope.
The first electron micrograph of virus particles was made in 1938 and belonged to a virus
from the poxvirus family called ectromelia, which was viewed directly in the lymph fluids
of an infected mouse (1). Amidst the microbial clutter, a number of suspected symmetrical
'viral' particles can be seen. After filtering and purifying further from the same lymph,
individual orthopoxviruses were revealed as grey ovoid blobs (2), but still the details of the
virus could not be made out. A 2012 image of ectromelia virus (3) magnified 20,000x
shows its characteristic rice-grain-shaped tubules.*

In 1946, Wendell Meredith Stanley won the Nobel Prize in chemistry (along with two other chemists) for the 'preparation of enzymes and virus proteins in a pure form'. A decade or so later, questions were raised about Stanley's 'crystals'. They were thought to be contaminated, and if so, his conclusions would have been based on a flawed premise. Nevertheless, Stanley's work showing the TMV's ability to turn into crystals and stay infective proved that viruses had a regular structure and were chemically stable.

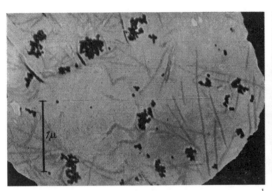 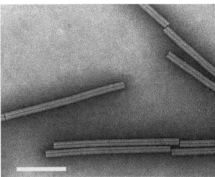

1   2

For nearly fifty years, TMV was known as an agent of disease long before it could be seen under an electron microscope. TMV was relatively simple to purify from an infected plant's sap and was isolated by crystallization in 1935. In 1938, Helmut Ruska in Berlin observed and studied the virus's structure, shape and size (1). Given its wide prevalence, ease of isolation and large size, TMV became a model for virus research for biologists. A 2012 negative contrast electron micrograph of TMV particle (2) shows its coiled cylindrical structure clearly (the white bar represents 100 nm, or 1/10,000, the width of a human hair).

Beijerinck didn't live to see his theory validated, nor did Pasteur get to see the bullet-shaped rabies virus, but scientists now had incontrovertible visual proof of life forms a whole lot smaller than the *smallest* bacteria. The images showed that viruses are simple but with sophisticated structures, each with the very function that made it infective. The advancements in microscopy became a turning point in the scientific understanding of viruses and disease and our bodies' response to them. The microscope is the first step for identification and validation of any microbe. It is invaluable in our fight against unknown diseases and vital in the rapid identification of agents, especially in events like bioterror, accidental leaks and spillovers.

With the arrival of powerful optical tools, a whole new world now opened up to science's inquisitive gaze. The discipline of virology was born out of bacteriology. Just a few months after witnessing the TMV under the electron microscope, the Berlin Group chanced upon some viruses that were in the act of killing a bacterium, which suggested that these life forms played a far greater role in nature than anyone had imagined.

The quest to make even more powerful microscopes in order to find even smaller infective particles continues to this day. There is an unknown and unending world of viruses and other minutiae out there waiting to be discovered, as is the possibility of gazing at the infinite.

Expect wonders.

SUPERSIZE ME

It is widely agreed that major evolutionary leaps—including the emergence of life itself—occurred in deep water or at the water's edge. Experts tell us that the first cellular life was jump-started in a tidal pool or in a deep ocean vent where energy and nutrients were mixed. And wherever there are cells and microbes, there are viruses! Water is teeming with an unfathomable abundance of these—from the salt pans of Qatar, the hot springs of New Zealand, Bengal's tidal mangroves, deep-sea vents in the North Atlantic Ridge, under an ice shelf in Antarctica or in a sewage drain near you—the diversity of viruses is outmatched only by their sheer numbers.

In 1992, Timothy Rowbotham, a microbiologist with the UK's Public Health Laboratory Service, was tasked with finding the source of a particularly nasty outbreak of pneumonia in Bradford, an industrial-age mill town in Yorkshire (made famous by Jack the Ripper). Rowbotham's detective work led him to sample water at the base of a hospital cooling tower. When he took his samples back to the lab and looked at them under a low-powered microscope, he could see amoeba. No surprises there. But the puzzling thing he noticed was that most amoeba had a large disc at their centre, which meant that they were likely to be infected by another microbe. Rowbotham named the microbe *Bradfordcoccus*—possibly one of the least appealing names ever given to a life form! He believed it was bacteria that had infected the amoeba, and for the next four years, he tried to find out more about the mysterious dark disc. Eventually, facing both a time and budget crunch, he was forced to put the unidentified microbe into deep freeze, but not before offering samples of *Bradfordcoccus* to French colleagues at the Aix-Marseille University. This centre was headed by a highly respected expert on infectious diseases, Didier Raoult, an imposing figure with long silver hair known for his unconventional, perhaps even eccentric, methods of research. Dr Raoult did not get a chance to look at the freeze-dried sample for some time. Then, in 1996, Raoult remembered this unidentified microbe and took it out of the deep freeze. What he saw under an optical microscope seemed to confirm Rowbotham's assessment—the dark disk certainly looked like a bacterium inside an amoeba. The sample also passed the Gram stain test, which is the first standard test used to identify bacteria.

This staining procedure had been developed way back in 1884 in Berlin by a Danish bacteriologist named Hans Christian Gram. It involves placing the wet streak of a bacterium on a glass slide and heating it gently until the liquid smear dries—a process called 'heat-fixing'. Over this smear on the slide, three dyes—crystal violet, iodine and safranin—are poured in sequence at intervals of two to three minutes. The smear on the slide does not look like much after this procedure, until you look at it under a simple microscope. If you see specks of pink, then you have a

 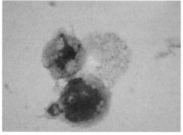 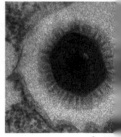

1   2   3

In 1992, while looking for the source of the Bradford pneumonia outbreak in the cooling tower, Timothy Rowbotham (1) found what resembled a round gram-positive bacteria with large blue vacuoles (2). Eleven years later in France it was diagnosed as a virus which infected an amoeba, Acanthamoeba polyphaga, *and was labelled as a 'microbe-mimicking virus' or mimivirus (3). Although Rowbotham's sensational discovery opened up a window into the world of enigmatic giant viruses, his career in public health ended with compulsory early retirement in 1998, at the age of fifty. He published several papers in international journals during his career and even had a species of bacteria (*Legionella rowbothamii*) named in his honour. He now runs a DIY hardware store in Bradford.*

Gram-negative bacterium like *Escherichia coli*, a common gut bacterium. Purple spots on the slide indicate it is a Gram-positive bacterium like *Streptococcus* that causes a sore throat and fever in humans. Gram-positive bacteria have thicker cell walls with a chemical make-up that allows them to retain the purple dye. In Gram-negative bacteria, on the other hand, the alcohol washes away the purple dye, and only the pink dye that is applied at the end gets fixed on these comparatively thinner-walled microbes. For pathologists, Gram staining was once the most useful tool for diagnosis, but with the rise of broad-spectrum antibiotics to treat suspected infections, its utility has diminished.

Rowbotham's sample came up purple, which showed that *Bradfordcoccus* was a Gram-positive bacterium. Raoult tasked Bernard La Scola, a talented bacteriologist in his group, to identify the bacterium's signature molecules such as ribosomal RNA (which helps the bacterium make proteins) to help classify it. But even after spending long hours looking for it, La Scola was unable to find any signature molecules in his sample. He then decided to place the sample under an electron microscope which was a 1000 times more powerful than his standard optical microscope. And there it was—the enigmatic creature in all its mysterious glory!

Bradfordcoccus did not have the smooth surface of spherical bacteria. Instead, it looked more like a soccer ball made up of many interlocking plates. La Scola now understood why he had not found the signature bacterial RNA in the sample—the microbe under the lens was not a bacterium at all. It was a *giant* virus with long fibres radiating out of it, a bit like Dr Raoult himself on a bad hair day!

The team announced its discovery in *Science* in March 2003, christening the virus 'mimivirus' (thought to have been abbreviated from microbe mimicking virus, although Raoult later admitted that the name 'mimi' came from an amoeba character called Mimi whose stories Raoult's father had read to him when he was a child). The mimivirus is not only massive in size, it has a large genome to match. It possesses 1262 genes, compared to just 100 or so in a typical virus. Mimi's DNA is comparable to that of bacteria, and it carries numerous cellular enzymes that had not been found in any other viruses known until then.

After the discovery of Mimi, more outsized viruses were reported from other places. A giant virus labelled 'medusavirus' was found inside a primitive amoeba, called *Acanthamoeba*, that lives in a hot spring in Japan. A virus-uninfected *Acanthamoeba* is capable of causing a rare but severe infection of the eye and skin in humans, and if left untreated, it invades deeper into the central nervous system, leading to death. But when the medusavirus infects the amoeba, it—Medusa-like—causes the amoeba to transform from a large, dynamic, shape-shifting cell into a hard, little cyst-like circle. Scientists have still not figured out why this happens and how this inanimate amoeba hunts for food or sustains itself in such an ossified state. More studies showed that the medusavirus is perhaps 'more evolved' than other previously known viruses because it contains within it a full set of eukaryotic proteins (called 'histones'). Its histones are used for folding its DNA within the cell nucleus which are independent of the proteins from the cytoplasmic jelly of an amoeba. The genes of medusaviruses and their ability to produce proteins show that not all viruses are obligate parasites and that it is possible that there are more than a few exceptional viruses that are capable of almost free-living. Other giant viruses like the pithovirus and megavirus are much smaller than mimivirus, both in absolute and genome-size terms. Interestingly, more than 90 per cent of the giant virus genes are unique to them and are not shared with any other microbes, strengthening the case for assigning them a fourth domain, distinct from bacteria, archaea and eukaryotes.

Some viruses are so large they can be parasitised by smaller viruses (called 'virophages'). The *Cafeteria roenbergensis* virus is a giant that infects tiny marine zooplankton, primarily drifting animals like krill and jellyfish. Marine biologist Tom Fenchel, who first described this virus, named it *Cafeteria* 'because of its voracious and indiscriminate appetite, after many dinner discussions in the local cafeteria [by the research team]'. Its specific name *'roenbergensis'* was prompted by a pink neon sign affixed to a wall on their residence in Roenbjerg, Denmark, which lit up just as the authors were about to give up on finding a good name for it. CroV, as the virus is more manageably called, is one of the largest viruses known. About half its genes are similar to others within its genome, but the other half, remarkably, has been borrowed from a wide range of bacteria, archaea, eukaryotes, and even other viruses. In effect, CroV has acquired genes from all the domains of life. CroV is one of the most complex viruses known, and it is able to synthesise numerous specialised proteins, which is an extremely rare ability among viruses. Yet, despite its size and complexity, CroV can be invaded by a tiny virus labelled *Mavirus* (short for 'maverick virus') that parasitises it. *Mavirus* is naturally present in zooplankton or is smuggled in through its cell wall to infect the CroV. When *Mavirus* integrates into the genome of cells of *Cafeteria roenbergensis*, it confers immunity to the host from CroV, and eventually kills CroV without causing any harm to the zooplankton cell.

After being a hot topic of conversation in conferences and cafeterias, giant viruses have been discovered in the gut of healthy and sick people alike. A recent study of the human gut suggests that we are potentially home to at least a few dozens different types of giant viruses, classified into seven different lineages, including two new families. The most common among these are mimivirus and marseillevirus. In 2014, a mimivirus was even discovered in a fourteenth-century coprolite (a polite term for fossilised faeces) found at a medieval site in Belgium.

The discovery of large viruses led to the reopening of the debate on exactly how viruses came to exist. Like every discipline of science that has at least one unending and vexing problem, the origin of viruses is *the* hotly debated topic among virologists at conferences and in pubs. Obviously, several theories were propounded about it, from dubbing viruses life forms that have come from outer space to calling them a product of spontaneous creation. Most of these theories, however, were too flimsy to stand up to rigorous scientific scrutiny and have mostly been discarded. But some theories have stood the test of time, and depending on their view about the origin of viruses, virologists now get slotted into two broad but opposing schools of thought. The first school believes that viruses evolved just before, or at least at the same time as when, life itself was emerging. This theory, called the *virus first hypothesis*, is among the earliest hypotheses and still stands, although it may be on its last legs. Its proponents hold that viruses emerged when a primordial soup of amino acids, lipids and small segments of RNA created the first cells. It was these early viruses that enabled the creation of new life by shuffling genes between cells and creating DNA from RNA, while providing cells with sophisticated machinery to generate their own energy, as well as the ability to communicate with one another. They became the precursors of archaea and bacteria, and since then, the virus-driven evolutionary engine has been working ceaselessly to create new life forms.

The opposing camp believes that viruses could only have arisen when cellular life was already in place. This school of thought splits into two, each propounding a slightly different mechanism of how viruses evolved. The advocates of the *escape (or progressive) hypothesis* believe that viruses emerged when genetic elements gained the ability to move between cells and these excised fragments of genes were surrounded by a cell membrane, and that gave rise to protoviruses. Such viral particles became a via media for communication between the earliest life forms like archaea and early bacteria, and gradually a few acquired the ability to infect cells and become obligate parasites. Genetic studies have confirmed that all creatures have within them such mobile genetic elements (MGE, also called 'retrotransposons' or, more descriptively, 'jumping genes') that can move back and forth within a genome. MGEs are an open-access system of genes and their exchange is facilitated by bot-like viruses who in deep time have accreted in each species to create more complex life. Cells of every life form have learnt, co-opted and passed down a variety of shared functions. MGEs make up an astonishing 42 per cent of the human genome, nearly half the entire genome of all mammals, roughly 70 per cent of some plants and up to 30 per cent of bacterial genomes.

The other sub-hypothesis contends that viruses are organisms that reduced their genome size and cellular functions and became dependent on hosts for their survival. This is labelled the *reduction (or regressive) hypothesis*. Medusaviruses and amoebae may have traded genes on multiple occasions in the deep past, and it was perhaps among the earliest viruses to have undergone reductive evolution. Medusavirus is placed under a new 'family' of giant viruses (Medusaviridae). Such downsizing can also explain why giants like Mimi and a few other large and ancient viruses like the herpes virus may have evolved from bacteria and then become enduring parasites in related hosts.

Most evolutionary biologists have now more or less abandoned the virus-first hypothesis altogether because it is unusual for parasites to *precede* the emergence of their hosts. Adherents of the reduction hypotheses argue that viruses should be made part of the (cellular) family tree, while promoters of the escape hypothesis imagine viruses to be squid-like bots from *The Matrix,* lifeless but infective and programmed to destroy, and that therefore, they have no rightful place on the tree. Some virologists have labelled viruses 'replicators' because they behave like perpetual machines that constantly shuffle and tinker with genes. They are blind, are not driven towards any goal and, as far as we know, work in a completely random fashion, driving evolution in no specific direction. These replicators take every opportunity to absorb and affix information in the form of genes from their hosts. A bit like bots or the machine learning algorithms of your search engine or social media which store all your data and use it to constantly feed you with information, some useful, much of it junk. This exchange of genes between viruses and the cells they infect

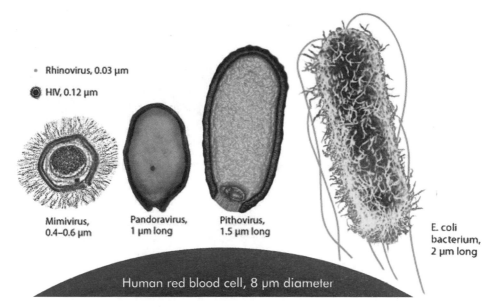

- Rhinovirus, 0.03 μm
- HIV, 0.12 μm

Mimivirus,
0.4–0.6 μm

Pandoravirus,
1 μm long

Pithovirus,
1.5 μm long

E. coli
bacterium,
2 μm long

Human red blood cell, 8 μm diameter

The size, shape and type of the genome of viruses can give us partial insights into how infective they are or can become. A study found that, in general, small-sized viruses cause more outbreaks than large-sized viruses, while some large-sized viruses are associated with a high case-fatality rate. The shape and size of viruses also determine their possible transmission routes. Giant viruses discovered so far infect other microbes, chiefly amoeba. The type of genome (RNA or DNA, single- or double-stranded) and their size also play a role in how infective viruses can become. RNA viruses have a higher probability than DNA viruses to find new host species because of their exceptionally short generation time and their faster, more error-prone replication cycles. With the exception of a handful of pox and herpes viruses, DNA viruses largely consist of viruses that have probably been present within or have co-evolved with human subpopulations for long periods of time. Most RNA viruses, on the other hand, have crossed over recently from animal hosts into humans. Of the 219 known human viral pathogens, 184 are RNA viruses. On an average, two new species of viruses are added to the list of potential human pathogens each year.

occurs both ways and both amalgamate each other's genes. This is what the arms race is all about in nature. Just a few genes decide which species will die out and which ones will survive. This ability of the viruses makes them one of the most persistent and potent forces of evolution.

The discovery of large viruses (termed nucleocytoplasmic large DNA viruses or NCLDVs) has brought some agreement among virologists that these may have evolved from eukaryotic cells, and it is probably true to say now that the balance has tilted in favour of adherents of the reduction hypothesis.

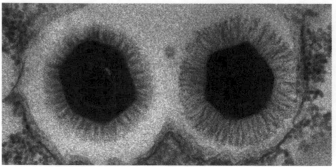
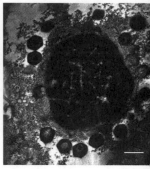

1

2

It is rare to find two or more viral infections occurring simultaneously. In exceptional circumstances, however, it is possible, especially when one gets immune-compromised through an infection like HIV or human herpesvirus infection, or when a patient takes immunosuppressive medicines before an organ transplant. In nature, a few viruses, like mimivirus (right, long hair [1]) and megavirus (left, short hair [1]) are known to work in partnership to cause simultaneous infection. Both of these infect and reside in the same vacuole of an amoeba. About seven hours after the infection, the megavirus forms a near-spherical compartment about the size of a large bacterium, and from its periphery bud out viral particles that soon leave the amoeba's body. Around five hours later, mimiviruses begin to bud and are released in a similar fashion. More megaviruses (about double) are produced than mimiviruses, and the reason why one is more fecund or numerous than the other is not fully understood. Scientists call this complex reproduction a 'virion factory' (2) and believe that such co-infections can also facilitate exchange of genes and thus further spur evolution. The scale bar corresponds to half a micron (the breadth of a human hair is 50 microns).

While the timing and origin of viruses may be contentious, there is little argument about how they evolved from ancestral viruses. This process is akin to the formation of an ancient, mighty river. It begins as a tiny trickle, which becomes streams, which in turn meet to form a rivulet. The quantum of water represents the varieties of genes, and the length of the river and its streams denote time. Several streams that join together bring discrete sets of genes, some ancient and related, others distinct and only recently evolved. As more streams join a single larger river, the flow of genes gathers momentum and this creates greater opportunities for the mixing of genes. As they flow, these genes become radically different from the ones that started as the original 'ancestor' trickle. Gradually, with obstacles and shifts, the river breaks up into tributaries which then diverge and go their separate ways. A few strands recombine to create a braided river, while others break away until they encounter a barrier, only to join again a little later. All this tumultuous wandering adds to the richness of new genes. Some gene flows are rapid and progress away from the main river. A few meet a dead end and become extinct, while others are stranded like oxbow lakes but still persist, and a few others dive under rocks and sediments and aestivate only to emerge when they eventually find an opening. The delta of the river is where all viruses flow unhindered into a giant sea, which is the proverbial virosphere. This churn of genes creates infinite opportunities for new viruses and viral particles to evolve and pass through diverse life forms. Thus, over several trillions of generations, once-cousins create progenies that are progressively more distinct from one another. This unceasing exchange of genes between viruses and their hosts not only creates new varieties that lead to new species but also helps to make major evolutionary leaps. By studying coat proteins and genes of viruses that infect algae, for example, scientists have identified the processes of how viruses enabled sea-living charophyte algae to become the earliest land-colonizing plants, a process that occurred around 470 million years ago.

Yet, we still do not completely understand why viruses have so few genes in common with other viruses. Even related viruses over time retain few genes that are similar. Had the earliest viruses emerged from a single or a set of ancient ancestors, they would have had shared genes that would appear across diverse viral groups, but they don't. In fact, many extant viruses are of quite recent origin. A few viruses that cause disease in humans have reached an evolutionary dead end, such as smallpox, polio and HIV. HIV-1, which originated from a chimpanzee lentivirus, underwent multiple cross-species transmissions before it jumped into humans in the late nineteenth century in a place somewhere between Gabon and the Congo. Once it had caused sufficient infections in people there, HIV made small genomic changes that increased its ability to transmit from one human host to another. This enabled it to spread out of Africa in the 1970s into the US and Europe, before it became a pandemic. According to a June 2021 paper published in *Current Biology,* humans may have encountered a coronavirus epidemic 20,000 years ago which occurred in East Asia. Gradually this virus down-mutated and caused a less severe form of disease. Another study has estimated that the lineage from which SARS-CoV-2 emerged had diverged from bats and pangolins more recently, just 800 or so years ago (or around 1200 CE), and became capable of infecting humans in the middle of the twentieth century. This means that the precursors to SARS-CoV-2 may have been residing unnoticed in bats and perhaps other mammals for decades.

Definitive answers to the origins of viruses will not be easy to arrive at and may never be conclusive. We will need to sift and combine elements from different hypotheses to arrive at a consensus, and in doing so, we will probably be delving into the very definition of life. Most scientists agree that viruses were not created in a single linear pathway. With each new discovery, viruses have even challenged our our definition and understanding of what exactly they are. But whether they are thought of as living or dead, or as the living dead, they often have transformative effects upon those who become their prey. As an interloper, a virus can turn the ordinary into a thing of beauty, cause sickness and death, and over generations, drive the process of evolution.

How can a 'dead particle' be so powerful?

4
THE VIRUS
IS US

The influential British biologist and science writer, Sir Peter Medawar, called the virus 'a piece of bad news wrapped in protein'. Scientists, too, can often be faulted for looking at viruses through the narrow lens of the infections and diseases they cause. But as the world's most abundant 'life forms', the ambit and influence of viruses is a whole lot broader.

Viruses are everywhere, infecting all forms of life, and it is this ability to shuttle and shuffle genes that makes viruses one of the most persistent and potent forces of evolution. Everything about viruses is extreme, including perhaps, the reactions they evoke. But for every truism about viruses, the opposite is also often true. Take for example, the speed with which they multiply within a host. While a single bacterial cell divides every few minutes or hours, viruses, too, can take just a few minutes or hours, like the viruses that bring about conjunctivitis or the common cold; or they can take several days, like the dengue virus; and a small minority can even do a Rip Van Winkle, lying dormant for years, like HIV does; or even for *decades*, as in the case of shingles.

Whether we see viruses as dead or alive, as life-threatening or life-affirming, there is an ineluctable beauty and even a certain elegance in the way viruses go about their lives. Each virus, under different conditions, adopts a unique strategy to multiply. Outside a host's body, a virus is inert. To spring back to life, it needs to inject its tiny package of genes into the right kind of cell, and in order to do that, it first needs to overcome a maze of complex defences of a host, locate the right cell inside the host, inject its genes into this cell, and only then is it able to make copies of itself, and hopefully, thrive. Whether it is their cunning in dodging a host's immune system, their craftiness in dismantling a host's DNA or their brilliance in making new copies of themselves—there is a majesty in the entire process which begins when the host cell starts to die until the emergence of several tens or hundreds of viruses. Every virus has a rhapsodic cycle of its own, and if you were to watch the viral invasion of cells under an advanced optical instrument, you might even hear in your mind's eye a march by Strauss, the brass band of Grieg or a Sibelius drama with drums rolling!

The first barrier that a virus has to overcome when invading a cell is its outer layer or membrane. This outer layer of a host cell employs a variety of gatekeepers and sentries to prevent the entry of invaders. Viruses, therefore, have to either trick these gatekeepers by making the host cell believe that they are one of them or by pretending that they are food for the cell and therefore need to be ingested. How do they do this against primed and alert host cells? Some viruses steal the host's membrane and develop a lipid (a short-chained fatty acid) envelope to surround their own protein capsid. This envelope, which matches the host's surface lipid or protein spikes, tricks the host into treating the virus not as an interloper but as a cell very similar to itself, therefore allowing it ingress. Once inside, the virus takes over the genome of its host and begins to replicate new strands of its own gene using the host's proteins and sugars. Soon, an army of mini-me is ready, all nearly identical to each other and all primed to take over more host cells. Each newly minted virus, just before it bursts out, acquires its own envelope—a flimsy, sticky mackintosh made of the cell membrane of the host, worn over its own protein coat—which helps it deceive other host cells into allowing access in subsequent invasions. After the viruses burst out, the host cell dies. This boom-and-bust strategy is called the 'lytic cycle' (Latin; *lysis*: to break).

Other viruses, instead of this wham-bam approach, adopt a more laid-back style. When the virus finds the right cell in the right host, it uses a protein key to open a protein lock on the surface of the host's cell to sneak inside. Once the cell wall is breached, the viral genome uses the host's enzymes to dismantle the host's genome and make copies of itself.

The viral genes then entwine and integrate with the host's genome, after which the newly embedded viral genes can remain ensconced inside the host cell for a very long period, often not doing much. The host cell, unaware of the presence of the virus's genome, goes about its business as usual. At some point, if the host cell encounters some form of stress, often, for reasons not known, the viral genome breaks free and resumes making copies of itself, using the host's DNA. Parallelly, it also begins to mass-produce coats stitched from the protein acquired from the host's cytoplasm (a jelly-like substance that separates the cell's outer layers and the nucleus). Eventually, these protein-coated viruses exit the cell to find and invade another host. This strategy of first merging and then breaking out is called 'lysogeny' (*lyso*: break; *-geny*: to produce).

Some viruses go a step further and adopt a rather extreme form of lysogeny. Using a Rip Van Winkle strategy, they insert their genes into the host's DNA and can then remain there, completely inactive or latent and undetected by the host genome for years or even persist across generations before 'waking up'. These viruses typically infect a sperm or

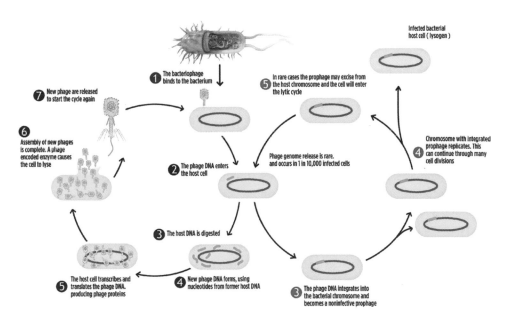

Infected bacterial host cell (lysogen)

① The bacteriophage binds to the bacterium

⑤ In rare cases the prophage may excise from the host chromosome and the cell will enter the lytic cycle

⑦ New phage are released to start the cycle again

⑥ Assembly of new phages is complete. A phage encoded enzyme causes the cell to lyse

Chromosome with integrated prophage replicates. This ④ can continue through many cell divisions

② The phage DNA enters the host cell

Phage genome release is rare. and occurs in 1 in 10,000 infected cells

③ The host DNA is digested

⑤ The host cell transcribes and translates the phage DNA. producing phage proteins

④ New phage DNA forms, using nucleotides from former host DNA

③ The phage DNA integrates into the bacterial chromosome and becomes a noninfective prophage

Two strategies for infection are adopted by viruses. The most common viral infections that make us ill are due to the aggressive lytic cycle of reproduction (3–7). However, all creatures also harbour viral genes, or entire viruses that lie embedded in their genome and live a passive life until they are provoked to cause infection. These viruses whose genes get implanted within the host's genome (through a lysogenic cycle) get stirred by an external stimulus and are released. This kickstarts a normal lytic cycle (3–5).

the egg cell of a host. Once they fuse, these newly inserted viral genes are copied into every single cell of its progeny, ready to be passed down to future generations. Some viruses that become long-term residents are actually good for their hosts. So good, in fact, that not only will the hosts be unable to survive without them, but entire ecosystems could collapse in their absence. We can go so far as to say that we humans may not even have come into existence without viruses. This astonishing assertion needs an explanation.

In 2003, we achieved a remarkable goal. The Human Genome Project, a collaboration between geneticists from twenty institutions from six countries, successfully mapped the *entire* human genome, enabling us, for the first time, to see what our entire genetic structure looks like. Decoding the human genome over the years led to a dramatic discovery about viruses. Scientists found that there are entire sequences of RNA viruses called 'retroviruses' embedded within our genes. Retroviruses are simple packages of genetic material that can implant their genome into their hosts' DNA. They convert their RNA into DNA using a special enzyme (reverse 'transcriptase') and are so called because this conversion of RNA to DNA is the reverse (retro) of the normal cellular process of converting DNA to RNA.

Although the first retroviruses were discovered at the turn of the twentieth century and the first human retrovirus in the 1980s, research into retroviruses accelerated only in the 1990s in an effort to understand one of the most insidious retroviruses of all—Human Immunodeficiency Virus (HIV)—and to find out how HIV leads to Acquired Immunodeficiency Syndrome (AIDS). Scientists discovered that HIV enters white blood cells and uses its RNA to make copies from the cell's DNA, thereby compromising immunity. There it can lie dormant for a long time, but once activated, infection sets in in the form of AIDS, which leads to reduced immunity that then makes the body susceptible to other diseases like tuberculosis (TB). HIV does not fully integrate into all our DNA and remains ready to infect other cells. Retroviruses like HIV are called exogenous (*exo-*: out; *-genous*: grow) viruses.

But the retroviruses discovered during the Human Genome Project were endogenous i.e. retroviruses whose viral genes had become part of our DNA. Virologist Robin Weiss first discovered an endogenous retrovirus in the 1960s when he discovered that jungle fowl and chicken genomes contained resident retroviruses that they had inherited from a common ancestor. Weiss's original manuscript that presented this interpretation in 1968 was at first rejected out of hand. Reverse transcriptase (the enzyme that converts RNA into DNA *inside* a host's cell) had not yet been discovered, and therefore, the process was thought to be 'impossible', as one reviewer put it bluntly. By the time Weiss's manuscript was published in another journal, other researchers had begun to report that they, too, had found retroviruses inside uninfected mice, pigs, ducks, monkeys and some other fowl as well, thus confirming the presence of endogenous retroviruses.

But what about humans? How and when did viruses get embedded in our genes? Studies suggest that the earliest acquisition and merger of an endogenous retrovirus (ERV) began when an early ancestor of mammals received a single retroviral gene that has stayed with all mammals ever since. Different mammals then acquired ERVs at different times in the course of their evolution. In the nearly two decades since the Human Genome Project, scientists have identified more than fifty distinct Human ERVs (or HERV) 'families' in human DNA. Of these, the HERV-L family is considered the oldest, and is estimated to have invaded the genome of an ancestor of all mammals some 150 million years ago. This was a momentous development for all modern mammals because it caused mammals to split into two distinct lines. The first constituted the ancestor of egg-laying mammals like the platypus and echidna. The other line consisted of mammals who give birth to live offspring ('viviparity') like almost all other mammals we are familiar with. Investigating the genes of animals that live today, palæobiologists and virologists estimate that the placenta in mammals was 'created' somewhere between 150 and 130 million years ago, after which most mammals took to giving live birth, using an ancestral form of the placenta which, too, had evolved largely through the action of ERVs.

The placenta is an amazing organ in more ways than one. It is a lifeline that connects a mother and her foetus, but it also forms a defensive barrier between them. The placenta provides the baby with oxygen and nutrients, carries away waste, filters out harmful microbes and pumps out hormones. And it does all this while keeping maternal and foetal blood supplies completely separate. The human placenta is a blend of maternal and foetal tissue. The way it forms is incredibly invasive. After the embryo implants into the womb, fingerlike projections burrow into the maternal tissue and alter its blood vessels so that they can remain bathed in a constant supply of the mother's blood. This interface, which has a surface area of 12 square metres, is what allows a mother and foetus to exchange nutrients and waste. Such close contact means that

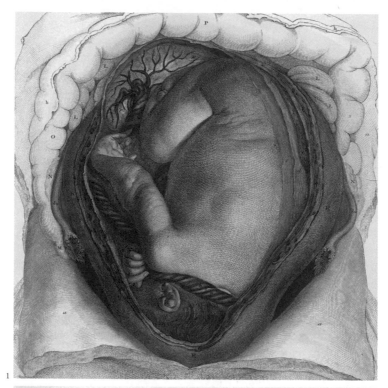

1

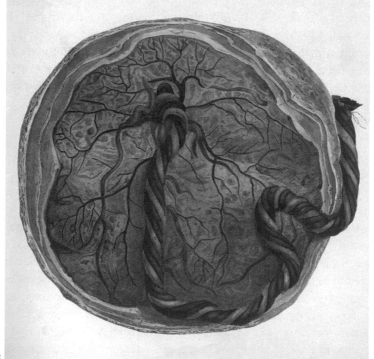

2

the mother's immune system can attack the developing embryo, which it can mistake as a 'foreign' invader. And this does sometimes happen when two parents of specific blood groups and blood factors conceive; the mother's immune response rejects the developing embryo in what is called 'spontaneous abortion'. But such abortions are rare, thanks to a series of HERV infections in our reptilian-mammalian ancestry. As a first line of defence, foetal cells along the boundary fuse together using a protein called 'syncytin'. Its name comes from its capacity to cause cells (*cyt-*) to combine (*syn-*), and these genes produce the protein that fuses individual cells into a constantly multiplying mass of cells, forming the layers of tissue making our placenta. These rapidly replicating cells (called 'syncytiotrophoblasts', Greek; *trophos*: to feed; *blast*: germinating) form a loose sac that provides nutrients to the high-energy-demanding embryo. This fusing also removes any gaps where maternal white blood cells could squeeze through and launch an immune attack. It is a clever strategy, but we did not come up with it entirely on our own. We had help from our virus friends because syncytin was originally a *viral* protein. The virus used syncytin to fuse with host cells so it could infect them. Then, during an ancient viral infection, a virus inserted its envelope gene into the host's genome. That genome along with its viral infection persisted and probably got passed on from one generation to the next, but that gene may have lain dormant for generations. Gradually, it was repurposed by evolution to fuse cells together in the placenta.

The placenta, orchestrated by a viral gene, is a remarkable organ. As a pregnancy progresses, the placenta grows in size and adds new functions. It acts as a filter to excrete waste produced by the foetus; as an air diffusion chamber for it to breathe; as a source for nourishment; and a gland that modulates hormones for the development of the foetus (1). The blood of the foetus and mother pass through separate arteries in the placenta (2). The foetus is not entirely selfish. Some types of foetal cells travel across organs and reach the mother's tissues, mostly those underlying her skin, liver, kidney and bone marrow. The role of foetal cells is not fully understood but they perhaps help heal minor injuries or diseases that the mother may encounter. Some studies have found that pregnant mothers make better survival decisions, and know when to fight, flee or fly. In every pregnancy, a mammal mother grows a new placenta to support her baby. Strangely enough, identical twins may or may not share the same placenta. This depends on whether or not the fertilised egg divides before it fixes to the uterine wall. If the egg separates before the formation of the placenta, there will be two and if it cleaves after, then just a single shared placenta will suffice. Once the pregnancy is over, the placenta is ejected out of the body, and it is perhaps the only organ that naturally expels itself when its purpose is fulfilled, quite like a virus. Some mammals consume their placenta after birth. This is because after an exhausting labour, the placenta offers an easy and rich source of iron, protein and vitamins for the mother. Physiologists have found that animals that consume the placenta immediately after birth are able to reduce post-delivery bleeding, restore their hormones quickly, increase their energy and, therefore, begin the supply of milk for their newborn.

In some ways, a developing foetus *is* a little bit like a virus, in that it exists inside the body of another organism where it needs to avoid detection and rejection by the (mother's) immune system. So perhaps it is fitting that it should be helped by syncytin, which is a protein synthesised from a gene taken from a virus. Apart from helping in the formation of the placenta, ERVs also helped regulate its nourishment and modulate the hormones that determine the gestation period and onset of labour. The gestation-regulating ERV inserted itself in the ancestors of Old World monkeys about 50 million years ago, and indirectly controls a hormone called corticotropin-releasing hormone (CRH) which regulates birth timing and pregnancy. HERVs also work closely with the mother's immune cells and protect the foetus from any processes that may adversely impact the development of the brain. So, it seems that an obligate parasite that gave us HERVs is an obligate requirement for human reproduction. Without it, we would perhaps still be laying eggs!

Another piece of the ERV gene that was inserted into the genome of a distant primate ancestor between 23 and 13 million years ago regulates the production of the enzyme amylase in our salivary glands as well as in the pancreas. This enabled our ancestors to thrive on a diet containing starch and eventually helped them make the transition from being hunter-gatherers to master farmers. Another set of HERVs regulate a cluster of genes encoding the globin (a protein) part of the oxygen-carrying haemoglobin in our red blood cells and are critical for modulating the switch from making the foetal to the adult form of globin during early development of the foetus.

HERVs have also been significantly involved in developing our limbs and our ability to move. Ancient, slightly curious viruses called 'foamy viruses'—so called because of the bubbly appearance of the cells they infect—began to incorporate themselves into ancient fish such as the coelacanth. These later infected other creatures like the mammals, especially primates and therefore us humans, as they evolved over time.

Understanding the complex mechanism and orchestration of ERVs and their host's genes is still a work in progress. For example, we are discovering that some ERV genes that have been passed down from ancient mammal ancestors have bestowed us with an inbuilt and inheritable defence mechanism called 'innate immunity'.

It is not all good news when it comes to HERVs, though. While our ancestors were still roaming the lowland forests and savannahs of Africa some 2,50,000 years ago, HERV-K HML-2 (HK2) became embedded in their genome. HERV-K are perhaps the most recent ERV insertions into our genome. Until quite recently, it was believed that HERV-K genes are not very active in the modern human genome but that may not quite be true. For example, HERV-K HML-2 (HK2) is unique to humans and is not found in any other primate, although other older HERV-K genes are present in all primates and have been passed down to us. This particular HERV meddles with mechanisms associated with pleasure in the brain and drives individuals to different types of addictions. Increasingly, this HERV and the proteins it produces is being incriminated for neurological disorders like multiple sclerosis, Lou Gehrig's disease, and even depression and schizophrenia. Can HERVs perhaps explain our Monday morning blues?

Some proteins associated with HERVs may even be responsible for triggering specific cancers (colorectal, pancreatic, liver, prostate, ovarian, breast cancer and melanomas) and inflammatory diseases. Thus, monitoring these genes and their proteins may provide early signals to

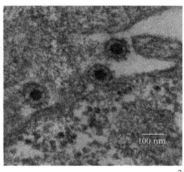

1   2

Hand-me-down: *The deep-sea dwelling coelacanth (1) has fleshy, lobed fins that look somewhat like limbs, as does the lungfish, an air-breathing freshwater fish. The coelacanth has lived in the seas for the last 410 million years or so. Believed by many to have no living relatives, the coelacanth is a truly unique species. They are the only living vertebrates with a jointed skull that swings upwards to greatly increase the gape of its mouth. Their limb-like pectoral fins are also unique, as they are internally supported by bone, a feature not found in any other fish. They use their fins in a paddle-like fashion, as if they are walking through the water column. Relict populations of this prehistoric fish lurk in submarine caves in the depths of the Indian Ocean. Nobody knows for sure why they need muscular fins, but some divers have noticed them 'walking' on walls of volcanic submarine caves and outcrops. It was about 400 million years or so ago that an ancestor of the coelacanth became infected by novel foamy viruses (2) which led to the development of its muscular fin, and that enabled an amphibious descendant to take the first steps out of water onto land. Most life forms—from deep-sea-dwelling coelacanths to chimpanzees—have their unique foamy virus infections. So you may have a series of viral infections in fish to thank for your dextrous fingers and toes, as well as for your mobile arms and legs.*

watch out for and to prevent the progress of these fatal ailments. Men need to be especially respectful of HERVs. A few specific HERVs are housed in the Y chromosome (males have X and Y chromosomes, women have X and X), and any aberration here causes male infertility.

Not all viral genes in the human genome are retroviruses. Some non-retroviruses, too, have been discovered to have taken up permanent residence in our genes and cells. After our *Homo sapiens* ancestors left Africa around 70,000 to 50,000 years ago, they encountered two of their cousins—the Neanderthals, who lived in the Steppes and Europe, and the Denisovans, who had left Africa about 5,00,000 years ago. We know from dating (of the molecular kind, not boy-meets-girl) of genes that Neanderthals and Denisovans have contributed about 4 to 7 per cent of the genes we possess today (in fact, a July 2021 paper has estimated that *only* 7 per cent of all our genes are unique to us!). Among them is a new strain of the (human) papilloma viruses (or HPV, which causes genital warts and can lead to cervical cancer) transmitted to us by Neanderthals via our ancestors. This could well be the first disease we humans acquired as a result of unprotected sex. In return, sapiens may have passed on some other infectious diseases to Neanderthals, such as helicobacter, which causes stomach ulcers, tapeworms and tuberculosis. It is possible that it was not just war and genocide but diseases contributed by our direct ancestors that led to the total extinction of Neanderthals.

Another non-retrovirus that is wedged between our genes is the bornavirus. Bornaviral disease, also known as the 'sad horse disease', was first noticed in horses in the German town of Borna. It affects nerves and the brains of its victims, often leading to their death. Bornavirus was the first non-retro RNA virus found in human DNA. This virus has also been found embedded in the genomes of other primates, as well as bats, elephants, fish, lemurs and rodents. No one knows exactly how they got there, but the timing of their evolution and divergence is intriguing. Modern bornaviruses emerged from a mammalian ancestor before the extinction of the dinosaurs, around 66 million years ago. These ancestors would have been small rodent-like mammals, the biggest perhaps as large as a capybara, scurrying terrified at the feet of the dinosaurs. Scientists believe that these were probably the first to include minuscule segments of bornaviruses in their genome. This was enough to gradually offer them some protection against the ancient bornaviral disease. They were perhaps assisted by a retrovirus that converted their RNA genome into DNA. In modern times, horses and sheep sometimes become infected, perhaps as a result of transmission by a shrew which acts as a reservoir. But it is

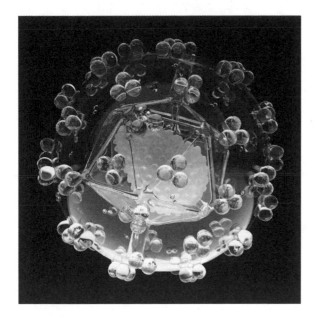

Papillomaviruses are DNA viruses and comprise several hundred species, each of which is specific to its host. Papillomaviridae infect virtually every vertebrate family, from fish to mammals. At least 170 different kinds of papillomavirus infections are known to occur in humans. Most don't produce any symptoms, some cause benign tumours or warts (called 'papillomas') and a few carry the risk of becoming full-blown cancers. Human papillomavirus (HPV, left) is the most common sexually transmitted infection and causes genital warts and can lead to cervical cancer.

reptiles and birds, who are descended from dinosaurs, that are still most affected by bornaviruses.

Not all ERVs take thousands or millions of years to be incorporated within a genome. One is taking place as I write, in plain sight, like a slowly approaching train. In 1988, a retrovirus was discovered that was infecting koalas (called the 'koala retrovirus' or KoRV) in Australia. KoRV makes this cuddly marsupial very sick and causes koala immune deficiency syndrome (KIDS) that is frighteningly similar to HIV/AIDS in humans. Once KoRV began to infect the koala germ line, it endogenised almost instantly, perhaps within a generation or two. An infected koala becomes lethargic as the debilitating leukaemia progresses and the koala eventually dies prematurely. While studying 120-year-old koala skins preserved in museums, virologists found remnants of KoRV genes, suggesting that the disease existed back then as well. So where did the virus come from? Virologists found that KoRV is closely related to the gibbon ape leukaemia virus (GALV). Gibbons are apes that live in the forests of South East Asia and in the north-eastern edge of India, too. Scientists believe that rodents hopped from one island to another to reach northern Australia. A chance encounter of a carrier rodent with a koala could have led to its crossover. Both GALV and KoRV are what virologists call an 'exogenous retrovirus', like HIV, which is the opposite of an ERV. KoRV is highly infectious and can be transmitted from one koala to another to cause disease. But scientists have found that when a

KoRV infection passes from a mother to its offspring (i.e., horizontally, instead of vertically) it gets incorporated within the offspring's genome like ERVs and causes mild or no disease. This ensures that the offspring survives. Geneticists guess that KoRV will gradually become an ERV in *all* koalas, but until that happens, KoRV will continue to kill koalas in large numbers.

Currently, there are only a few stable populations of koalas in pockets of northern and central Australia where KoRV has become incorporated as an ERV. As the virus moves southwards, it remains a serious threat. Only the island of Tasmania and some patches of forests in southern Australia remain unaffected by the disease. Some southern koalas which have been infected with the KoRV virus have shown greater resilience than their northern cousins. This is because the southern KoRV strain has a large disease-causing chunk of its genetic material missing from it. The larger and fluffier southern koalas are more inbred, which could be a reason why their genes have slowed the spread of the virus. Another recently endogenised ERV, the cervid endogenous retrovirus (CrERV) in mule deer, has a similar mechanism but scientists don't know as yet whether they are infectious or not (perhaps not, according to some experts).
Is it possible that like KoRV, HIV, too, will get incorporated into the human genome in the distant future? While it is not beyond the realms of possibility, virologists think it is unlikely. Human germ line cells are not 'permissive' to HIV. HIV belongs to a lentivirus family that endogenises readily in other mammals but so far has not been seen in the lineage of the great apes, including ourselves.

Geneticists have discovered that some HERVs can lie dormant for a very long time and are then awakened by environmental triggers (such as infection by other viruses, drugs or mutations), often with dangerous consequences. In 2006, researchers at the Institut Gustave Roussy in France extracted parts of HERV-K from the human genome and reconstructed it to what may have been the viral gene that invaded and inserted itself in deep time. They then introduced it into a human cell whose DNA did not contain one. They discovered that the HERV-K reinserted itself into the human DNA. When they brought HERV-K close to DNAs of hamsters and cats, it vigorously embedded itself in them. These experiments suggest that HERVs are still infectious and if they escape from our DNA, they can cause diseases or trigger other kinds of evolutionary changes that we cannot even imagine. Future research on the applications of HERVs must therefore proceed with utmost caution.

1   2   3

For scientists, the koala is a genetic conundrum. The ancestors of the modern koala, taxonomically labelled Phascolarctos cinereus *(Greek;* phaskolos: *leather bag or pouch;* arktos: *bear; Latin;* cinereus: *ash-coloured) emerged around 15 million years ago as tree-climbing rainforest-dwellers. Modern koalas comprise three subspecies: the Queensland koala (*P. c. adustus*) (1), the New South Wales koala (*P. c. cinereus*) (2), and the Victorian koala (*P. c. victor*) (3). These are distinguished by the colour and thickness of their fur, body size and skull shape. The Queensland koala is the smallest of the three, with shorter, silver fur and a shorter skull. The Victorian koala is the largest, with shaggier, brown fur and a wider skull. Geneticists and taxonomists disagree on whether or not koalas should be separated into subspecies, but most agree that their behaviour is distinct. Adult Queensland koala males, for example, are highly territorial while Victorian koala males are not. Genetic studies suggest that limited gene flows between populations because of geophysical barriers like forest breaks, rivers and even roads may have prevented intermixing, and the three 'subspecies' may have diverged not very long ago from a single ancestral population.*

There is a real-life example from the plant world that shows that ERVs liberating themselves and becoming infective is not wholly impossible. The banana streak virus (BSV) became integrated in a banana species as an ERV during several episodes of gene transfers. But researchers found that when a banana is under fungal or insect attack, the BSV genes free themselves to immunise the plant. This characteristic has also been seen in a different ERV of tomatoes. This apparently selfish viral element that is otherwise napping in the host's genome appears genie-like at the right time to protect it.

Virologists at Oxford University have reconstructed the history of our viral DNA and estimate that there were at least thirty to thirty-five separate invasions which embedded ERVs into our genome. The virus that endowed us with HERV-H (they are critical in the development and fusion of the sperm and egg cells and in subsequent cell development) was a super-spreader because it got transmitted through the germ line or from parent to child. HERV-H is the most persistent of all ERVs and is found across all simians and apes, but some ERVs were discarded over the course of evolution. There is no clear answer to why we have fewer ERVs than our ape and primate ancestors. But losing ERVs also increases our risks to invasion by other viruses, and this may explain the rising numbers of cancers or how exogenous (outside) residents like HIV became a persistent infection. There is no way of telling if we are done with acquiring any more ERVs for now, or whether some day HIV or another retrovirus will inveigle itself into our genes.

The growing understanding of HERVs' functions has enabled geneticists and doctors to counter problems that were previously considered intractable. In 2015, medical scientists at the University of Rome suppressed the overexpression of one particular HERV (HERV-H) in a twelve-year-old boy with attention deficit hyperactivity disorder (ADHD) using a drug called methylphenidate over a period of six months and mitigated his condition. And although it is still early days, researchers are looking into the possibility of harnessing the activation of some HERVs to reverse the action of cancer cells. This novel therapeutic approach is a promising area of research for the future where instead of passively monitoring HERV proteins, innate immune responses can be instigated to cure cancer.

ERVs will play a crucial role if regulators and ethicists approve the use of animal organs for human recipients (or xenotransplantation, *xeno*: stranger). Pigs are prime candidates for harvesting organs, and lungs from genetically modified pigs are remarkably similar to our own. The Chinese government appears to be aggressively funding research and trials in xenotransplantation. Some biologists, however, believe that there are inevitable risks of porcine endogenous retroviruses (PERVs) getting transferred into a recipient's tissues, bringing additional problems with it. Although there has been no reported case as yet of these weak virus crossovers even among immunosuppressed patients, science is quite rightly erring on the side of abundant caution.

When the Human Genome Project first confirmed the presence of HERVs, Robin Weiss of University College, London, who was among the first people to discover ERVs in wildfowl and domesticated chickens in the 1960s, told *The New Yorker*, 'If Charles Darwin reappeared today, he might be surprised to learn that humans have descended from viruses as well as from apes.' There is no ignoring the viruses within us. Viruses tinker with genes and shuffle them around as they go about their business, and we humans have perhaps been one of their greatest beneficiaries in terms of the sheer numbers of ERVs we have acquired. These viruses, wedged between our genes and those that lie outside our genome, protect us from other infections and assist in shaping the body plan of every new individual, from an embryonic stage till the end of our lives. Humans may owe their very evolution and survival to hitch-hiking and stowaway viruses. The study of HERVs and viral genes has shown that they are neither junk nor vestigial. We have thus far discovered only a few vital functions they perform; there are probably many more viruses as well as benefits that these genes provide that remain unknown to us. As we get to know more about other viral genetic sequences and discover new ones, we are likely to be in for more surprises and mind-boggling revelations.

As they say, watch this space!

5

A DEEP
CONTROL

It's not easy being tiny *and* a total parasite. Viruses spend a lifetime out of sight and virtually out of mind—blending into the air, buried in the soil, afloat in water or hitch-hiking on a host. But the self-effacing virus plays a disproportionately vital role on Earth by making all life possible. The complex network of life is driven by the action of microbes who simultaneously create, consume, break down, synthesise and fix organic matter. They also produce energy with relentless and tireless efficiency. At the heart of every critical process in the biosphere—from creating the atmosphere to decomposing waste—lie microbes, and maintaining control over these are ultimately, what else? Viruses!

After the Sun and plate tectonics (gargantuan geological processes by which the continents and oceans, landscapes and rocks are created or destroyed), viruses are the most significant force to shape the course of evolution on Earth. Viruses have the power to cause billions of infections every second and to reproduce within their hosts quickly. In doing so, they create opportunities to cause mutations which, in turn, create new varieties of themselves and that of the host they infect. It is in this sense that viruses are a powerful unseen force that keeps the planet and its complex systems in order or throws it into chaos. Geological processes for the most part take place at a very slow—geological—pace. Every day, the Sun's radiation and UV rays pour unlimited energy on the surface of the Earth. It is solar energy that powers wind patterns and ocean currents, shapes the seasons and determines climate. Viruses, on the other hand, work much faster, but their actions, like viruses themselves, are covert and little known.

More than two-thirds of the Earth's surface lies under the sea and the top 20 metres of all seas are dominated by photosynthetic microbes, in particular a variety of bacteria called cyanobacteria (Greek; *kuaneos*: dark blue; *baktēria*: staff, cane, presumably because the first ones to be discovered were rod-shaped). The earliest cyanobacteria arrived on the scene about 3.6 billion years ago, making them roughly three-quarters of the age of the Earth itself. Over 3 billion years or so, cyanobacteria, at first by themselves and later in partnership with other microbes, formed colonies at places where shallow seas met the land. These colonies were 1 to 4 feet high with rounded or cabbage-like shapes. They are called 'stromatolites' (Greek; *stroma*: stratum; *lithos*: rock), and each colony comprised billions and billions of photosynthetic cyanobacteria that developed around calcium-rich rocks like limestone. They were not much to look at, and resembled modern-day toadstools. For over 3 billion years, free-living cyanobacteria and other photosynthetic organisms in water, together with stromatolite colonies, were the *only* life forms on Earth, and it is these early microbes and their partnerships that produced *all* the oxygen we breathe today. They produced enough of it to saturate the atmosphere with oxygen, create the ozone layer which protects the earth from the ultraviolet rays of the Sun and helped to form new minerals on the surface of the Earth.

The cyanobacteria grew on top of the water, and to avoid being smothered by sand or sediment as well as to access as much light as possible, they also grew steadily outwards. As they grew, they added new layers over the inorganic calcium- and silica-rich sediment, upon which more cyanobacteria grew, and this process was repeated over and over again. When they died, cyanobacteria were immaculately preserved in layers of minerals that smothered and entombed them. When the first microscopic animals appeared on Earth around 600 and 580 million years ago, they preyed upon these photosynthesisers, and stromatolites gradually began to disappear, eaten out of existence by these new, rapidly multiplying mobile grazers.

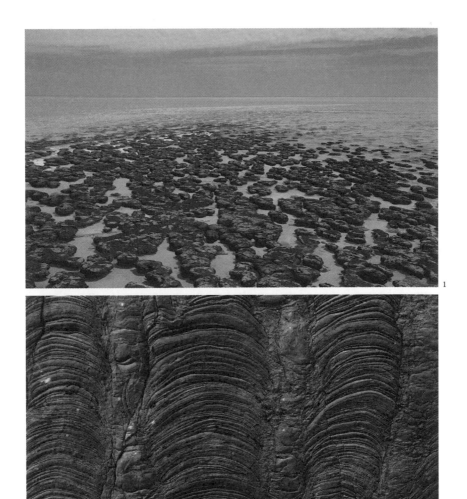

1

2

Hamelin Pool in Western Australia is home to the largest surviving stromatolite colony in the world, spread over 1270 square kilometres (in comparison, for example, the city of London is 1580 sq. km) (1), Each of these 'living rocks', resembling charred broccoli heads, is made from partnerships of two species of cyanobacteria and other microbes. For reasons not clearly understood, this stromatolite colony has stayed protected from the assault of microbial grazers like slugs and crabs, or overly aggressive phage epidemics that could have easily decimated them. Fossils of stromatolites when cut, like this 2.6-billion-year-old fossil from Chhattisgarh, India (2) show how microbial colonies were built layer upon layer, quite like the stromatolites that currently thrive in Hamelin Pool. Rising sea level and pollution can potentially introduce predators which will feed on them, or out-compete the stromatolites for nutrients. Stromatolite colonies need as much protection as we give to large, charismatic species. They can provide valuable insight into how we can mitigate climate change in the future using microbes.

Stromatolites are all but gone now, except in a few places like a lagoon in the Bahamas and Hamelin Pool in Western Australia, where they are still growing very slowly, adding about 5 centimetres every century. How they escaped the attention of voracious microbes that feed on them or grazers like crabs, slugs and other molluscs is something of a mystery. Some scientists believe it could be because the places where these stromatolites survive do not have grazers, and the conditions there are much like they were when these colonies first formed and flourished.

With the stromatolites all but gone, it is free-living cyanobacteria and other photosynthetic organisms that have been chipping away at carbon dioxide levels in the atmosphere and are producing most of the oxygen we breathe today. The dominant cyanobacteria of open seas are *Prochlorococcus*, and at just 0.5–0.7 micrometres (μm) in diameter—only a quarter as thick as the thinnest strand of a spider's web—they are among the smallest free-living photosynthetic organisms. If you don't consider viruses as living things, it is cyanobacteria like *Prochlorococcus* that are perhaps the most abundant creatures in the world. Which makes it rather interesting that this—the world's most abundant life form— remained undiscovered until 1988. Since then, however, several other abundant photosynthesisers have been discovered each year, resetting our understanding and deepening our appreciation for these ocean-dwelling creatures. *Prochlorococcus* (along with a distant cousin, *Synechococcus,* [Greek; *synechos*: in succession; *kokkos*: rounded granule]*;* and another ancient photosynthetic organism, SAR11, or the <u>Sar</u>gasso Sea <u>11</u>th DNA sequence) dominate the top layers of the world's seas and each of these bacteria has a clearly defined niche within different layers of open waters. The photosynthetic bacterial community collaborates with other microbes like sulphur-producing bacteria to produce nearly 70 per cent of all the Earth's oxygen and bury more than 4 gigatonnes of carbon dioxide (a gigatonne is a billion metric tonnes, GtC) every year. This is more than all trees and human efforts to scrub carbon achieve. In all, oceans sink 9 GtC while forests and vegetation on land absorb about 11 GtC. The important point here is that forests and vegetation on land are both a source and sink of carbon, while oceans are just sinks of carbon and net oxygen producers. So it is this green slime floating over ponds, lakes and wafting around in choppy seas that we have to thank for the air we breathe.

But there is a catch. Left to themselves, marine photosynthetic bacteria would multiply endlessly, turning the ocean into a slimy green, stinky pea soup, like sewage-filled rivers or drying ponds. This is where viruses come in. More specifically, viruses that infect bacteria which are called 'bacteriophages', or simply 'phages' (phage pronounced as faay-j). They play a critical role in the ocean's nutrient cycling. Bacteriophages called cyanophages (eaters of cyanobacteria) are the most abundant biological entities that feed on the most abundant bacteria, the cyanobacteria. Cyanophages' numbers are estimated to be up to 4×10^{30} or roughly 100 times more than all the microbes in the seas, and they mostly infect and kill mature cyanobacteria. The rate of viral infection in the oceans stands at 10^{23} infections per second, and these infections kill 20–40 per cent of all bacterial cells in the oceans *each day*. When cyanobacteria are invaded and killed by several viruses, their remains sink to the bottom of the ocean. Unlike trees, whose remains decompose on the surface of the land and release organic carbon dioxide into the atmosphere in a matter of a few weeks to a few years, killed cyanobacteria descend to the bottom of the oceans and remain buried in deep marine sediments for thousands

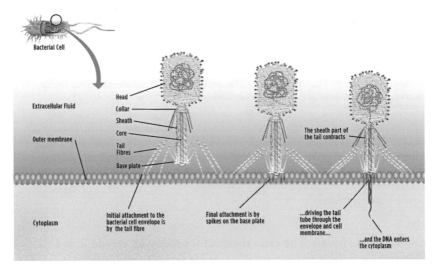

There is a phage for every bacterium, and this makes them the most diverse entity on the planet. Each phage is slightly different in shape and size, and each has its own idiosyncratic lifestyle. Phages have perfectly geometric designs and they are described based on the structure of their head which houses their genome. Despite being extremely tiny, the genome of phages have to withstand enormous pressure, up to 50 atmospheres or the same as being under 1800 feet of water at the equator. What protects them is a specially designed protein shell called a 'capsid', which is most commonly found in the form of an elongated icosahedron (a shape with twenty equilateral triangles, arranged together in symmetry) but there can be several variations of this complex three-dimensional structure. Some phages resemble the 1950s Russian Moon Landers. Attached to the head is a singular rod (the sheath) which ends in a spindly, daddy-long-leg-shaped tail. And it is the tail's design that make phages highly specific towards attaching on a bacterium's surface. The tail consists of a base plate, central tube, tip, sheath (which acts like a pilum which digs through the host cell's outer wall), a spike and tail fibres which when fixed on a bacterium's cell wall resembles an oil rig, and is used to insert its genome into the host cell.

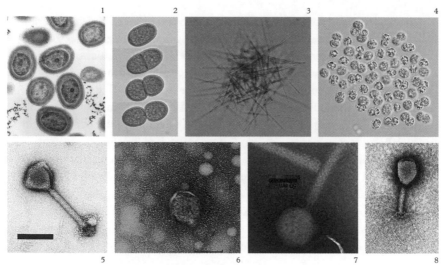

Those magnificent cyanobacteria *and their killing machines!*
Cyanobacteria began to dominate the water and moist soils from around 3.2 billion years
ago. They manufactured all the free, breathable oxygen we have today and created the ozone
layer. Among the 2000 species of cyanobacteria, the smallest known and most abundant is
Prochlorococcus *(1). Its cousin,* Synechococcus *(2) is an oval-shaped bacterium and*
is many times larger. Trichodesmium *(3) is perhaps the largest nitrogen-fixing microbe*
and recycles nearly half of the estimated marine nitrogen. If large quantities of human-
produced organic waste flow into the open sea during warm periods, Trichodesmium *and*
Microcystis *(4) are among the first cyanobacteria to begin forming colonies. To make the*
most of a nutrient rush, they morph into filamentous tuft or puff-like forms and harbour
a variety of bacteria within their strands. Their explosive growth can lead to algal blooms,
and over large areas of open seas, it disrupts the oxygen-carbon dioxide levels and suffocates
marine life. Below each cyanobacterium you can see their specific phage (in grey) which
regulates their population.

and millions of years. The crucial aspect for reducing carbon in the
atmosphere is deep burial, and the role of viruses is critical to this process.

Phage-infected bacteria absorb more carbon before they die and sink to
the ocean floor, compared to those that die uninfected, because when
cyanobacteria and other microbes are killed by bacteriophages, their cells
burst and release ions of carbonate that rapidly react with free calcium ions
in seawater to make calcium carbonate. All of this calcium carbonate and
the dead organic remains of microbes descends slowly into deep, dark water
like snowflakes, and can remain there for thousands of millions of years.

Viruses are incredible killing machines—according to one estimate,
phages can destroy an estimated half of the world's bacterial population
every forty-eight hours. If bacteriophages are so abundant and such

prolific killers, what prevents them from killing off *all* cyanobacteria? What regulates viral populations in the oceans, or anywhere else, for that matter? Until quite recently, few virologists could answer this question with any degree of confidence. Most assumed that phages drift about endlessly until they find a host cell to infect, and that phages have no known enemies that prey or feed on them. A definitive answer to this puzzle was provided by a study published in March 2020, when Dutch marine biologists observed sea-dwelling creatures like crabs, cockles, oysters and sponges in controlled tanks. For three hours, each individual creature was placed in a separate tank containing the same strength of viral populations. After three hours, the numbers of viruses were counted again. They found that sponges, which filter in and filter out specks of particles from seawater, are champions at ingesting large numbers of viruses. They removed up to 98 per cent of the viruses; crabs came second with 90 per cent reduction, followed by cockles which managed to trap 43 per cent, and oysters which captured just 12 per cent. Interestingly, we still don't know how viruses that we ingest are killed inside our bodies.

Present on the surfaces of oceans and seas, along with the cyanobacteria are free-floating photosynthetic organisms called 'plankton' (Greek; *planktos*: to wander or drift). Plankton are tiny creatures that live on or near the ocean's surface where there is enough light to allow photosynthesis (or phytoplanktons, *phyto-*: plant) or those that feast on the photosynthesisers (the zooplanktons, *zoo-*: animal). The outer layers of some plankton are made with carbon-and silica-rich shells and delicate skeletons. Like cyanobacteria, plankton too are primary producers, the foundation of the ocean's food pyramid. Plankton too are killed by viruses, and they add to the mass of organic matter sinking to the ocean floor. Over tens to hundreds of millions of years, plankton and cyanobacteria accumulate, forming reams of calcium-silica shells and organic carbon-rich cell matter that stack up to become layers of rock several hundred or even thousands of metres thick under the sea. Gradually, over millions of years, depending on the upheavals and subduction in the sea floor, these rocks are compressed under the weight of water. If the remains are chiefly calcium carbonate, they will form layers of chalk like the white cliffs of Dover and Calais. If the submarine chalk landscape is heated by roiling magma from below the crust and is pushed up, it creates marble crags like those in Carrara, Italy, from which Michelangelo hewed David. If copious amounts of organic matter are deposited and decompose in the absence of oxygen, over several millions of years, the ooze turns into various kinds of oil. Recent studies have even found a small community of petroleum-dwelling bacteria along with their specific bacteriophages.

A similar predatory relationship between different bacteria and their bacteriophages also helps in the capture and destruction of methane, the most potent natural greenhouse gas. Nearly 252 million years ago, the greatest extinction event called the 'Permian Event' that killed off nearly 95 per cent of all life on land and sea was caused by the sudden release of huge amounts of methane from a large underwater seam of clathrate which had been heated by volcanic or tectonic activity under the sea. Clathrate is a white, crystalline form of methane that forms and resides in deep seas. Methane is emitted by belching and farting cows, by the cultivation of rice, decaying organic urban waste, oil wells and pipelines, volcanic vents and of course, by ancient shallow marine deposits (clathrates)—not necessarily in that order of magnitude. In the oceans, methane-digesting bacteria are found concentrated around deep-sea vents and clathrate deposits, where they feed on leaking methane. These methane-gobbling bacteria generate energy for themselves by transforming methane and sulphate into carbonate and sulphide. A complex community of methane-dependent bacteria and their very specific viruses operate in a fashion similar to cyanobacteria and their viruses—bacteria devour the methane and are killed by phages on maturity, and then sink and are buried in the depths of the seas. Understanding the relationship of carbon dioxide and methane with bacteria that utilise and assimilate them, and finding out how viruses control their populations, can prove to be an important weapon in our fight to mitigate methane, and thereby, climate change.

Viruses and other microbes don't just control the metabolism of carbon on water and land, but also regulate phosphate, a key ingredient that makes energy transfer possible in cells. Viruses mediate the flow of phosphates within the microbial loop. They also help bacteria to assimilate and fix sulphur to make essential amino acids and enzymes. Together, carbon, phosphorus, nitrogen and sulphur are the building blocks of life, and viruses play a role in regulating the life cycle of each one of these elements. Viruses also help bacteria, algae and protozoans (those translucent, shape-shifting single-celled creatures like amoeba) to assimilate nitrogen which is integral to making proteins and amino acids in fish and other animals in the seas.

On a completely different tack, viruses that infect algae (Latin; *alga*: seaweed) create an interesting localised weather phenomenon. Algae are photosynthetic creatures that live in water or moist places and are closer to being considered plants than animals. They come in a wide variety of shapes and sizes. They can be minute and free-floating like plankton, or fixed and very plant-like and even massive like giant kelp and seaweed.

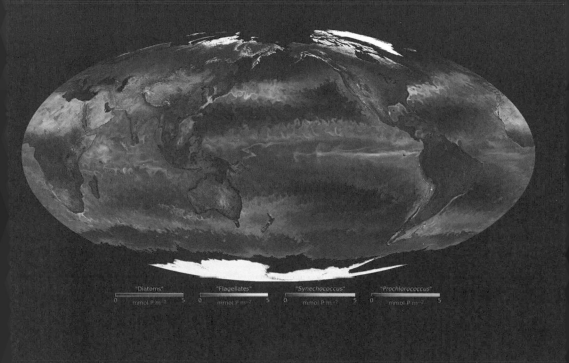

"Diatoms"   "Flagellates"   "Synechococcus"   "Prochlorococcus"

0   5   0   5   0   5   0   5
  mmol P m⁻²   mmol P m⁻²   mmol P m⁻²   mmol P m⁻²

Anywhere between 50 and 80 per cent of Earth's free breathable oxygen comes from oceans, thanks to micro-photosynthesisers. Among a variety of these primary producers, the champion cyanobacterium, Prochlorococcus, is the smallest photosynthetic organism on Earth. It alone produces up to 20 per cent of all oxygen, which is higher than all of the tropical rainforests put together. The rapid turnover or the birth and death of micro-photosynthesisers is calibrated by specific marine bacteriophages. The net photosynthesis that takes place in the oceans determines global oxygen availability and carbon-fixing. By using satellite imagery like this one by NASA, scientists have mapped where different marine photosynthesisers like drifting plants, cyanobacteria, diatoms and algae are concentrated. The highest photosynthetic activity is observed along the equator (marked with green) followed by eastern margins of coasts (blue and green) and high-latitude ocean where diatoms do most of the photosynthesis (red).

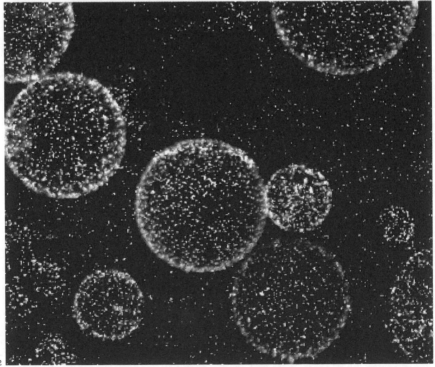

Algae lack true roots, leaves or fibres or highly differentiated tissues like true plants, and are very resistant to the action of bacteria in the sea. This is one reason why agar—a material extracted from red algae which grows in abundance in the Western Pacific Ocean—is infused with nutrients and used as a culture to grow bacteria in petri dishes in labs. Even though resistant to bacteria, algae are susceptible to a variety of viruses. They become infected by some specific viruses that manipulate algae to release a gas called dimethyl sulphide (DMS). You may have experienced DMS if you have travelled on a warm day or evening on a ferry. As you approached the wharf, you would have encountered a whiff of a tangy, almost sweet odour, a bit like the smell of boiling corn or beetroot, which sailors call 'the smell of the shore'. It smells different to the lungful you take at the beach, the salty and slightly sea-fishy smell. Seabirds are attracted to the smell of DMS and swoop down to where they find it strongest. Like all aromas, this is richly rewarding because this is where they find algae and plankton galore, which in turn, attract fish. These seabirds too contribute by fertilizing the water with their poop. DMS is found in higher concentrations in colder waters like in the bays of the Southern Ocean which collars Antarctica; Nova Scotia and the Grand Banks region; the area between the western Alaskan coast and Kamchatka; southern Chile and Argentina; and the vast expanse between Norway, Iceland and northern France. When DMS rises into the air, its molecules form an invisible, low-hanging cloud which attracts moisture in the air, creating a mild sea mist. As molecules aggregate, they begin to seed clouds, and if a lot of DMS is produced in the open seas, it contributes to more rain in the sea or along the shoreline. These low-hanging clouds also reflect back incoming sunlight into space, and this cools the planet as well. Clouds that rise above the Southern Ocean reflect

Some algae emit a chemical called dimethyl sulphide (DMS) which gives beaches a characteristic bracing, briny sea smell. This odour is a signal for fish that there is a concentration of algae and marine plankton feasting on nutrients. The aroma is irresistible for the fish and chasing them are birds. Next time when you fill your lungs with sea breeze and hear the cawing of seabirds, note the sulphurous hint of DMS in it. Scientists have observed that DMS is produced in cycles because marine phages, like janitors, work overtime, restricting the overpopulation of algae and other microbes in seas. If there is widespread death of DMS-producing microbes by excessive phage infection, other microbes pick the chemical warning and protect themselves against a possible viral infection. The DMS-emitting algae proliferate again only when nutrient concentration gets high and they 'sense' fewer phages around them. Often when these DMS emitters proliferate close to land, waves can carry a thick foam of DMS-emitting algae to the beach, like here along the Dutch North Sea coast (Dr Jacqueline Stefels, a plant ecophysiologist at the University of Groningen to scale) (1). This foam is produced by the algal bloom of Phaeocystis (2), which is a major DMS producer in cooler seas and plays an important role in deep marine carbon burial and the sulphur cycle.

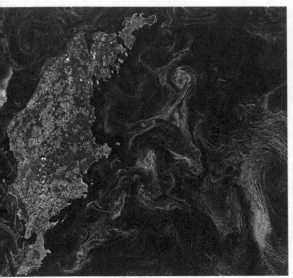
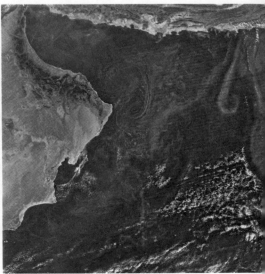

Algal blooms are not truly algal but are triggered by cyanobacteria partnering with other marine microbes. No two blooms are alike. A sudden nutrient flux and warm temperatures get several species for their population to explode with: A sudden nutrient flux and warm temperatures cause an explosion in the population of a number of species. Above, swirls of algal bloom spiralling across the Baltic Sea in the summer of 2015 (1); dust and sewage from large coastal cities of Pakistan and western India cause frequent algal blooms like this

more sunlight in summer than they would without algal and plankton growth. As rain falls on land, it absorbs the carbon dioxide present in the air, making rainwater mildly acidic—a very diluted version of the fizzy cola you drink. Scientists are exploring ways in which algal DMS can potentially be harnessed to absorb carbon dioxide, produce more rain in the oceans and help to mitigate global warming.

In some situations, certain kinds of marine microbes can become a nuisance. When massive organic pollution—such as fertiliser run-off from farms, or sewage—enters shallow seas, it causes an explosive and noxious growth called an 'algal bloom'. Here too, viruses come to the rescue and kill the excess microbes and restore a balance. But of course, it takes time for the viruses to overcome a massive algal goop. Left unchecked, algal blooms can deprive fish of oxygen and kill birds, poison the water and have a widespread impact on the fisheries industry.

The moral of the story, if we are to look for one, is this—in order to reverse the earth-warming effects of producing excessive carbon dioxide and methane, we need to bury vast amounts of carbon in the depths of

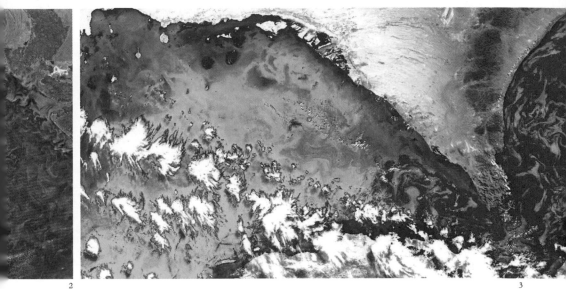

one in 2012 (2); and desalination plants are both the cause and victim of red tides like this one invading water-treatment facilities in the United Arab Emirates from 2008 (3). Phages can be deployed to correct ecological imbalances like algal blooms and locust swarms. Because of their specificity, phages kill specific species without affecting another, and after their job is done, they quietly fade away. Scientists are exploring ways by which they can artificially stimulate algal blooms and use specific phages to capture and sink carbon in deep seas.

the earth or seas. Moreover, carbon needs to *stay* buried and unperturbed for a very long time. This is precisely what happens when myriad bacteria, viruses and a multitude of other minutiae busily carry on their wild rumpus in the oceans, in open seas, lakes, rivers, ponds and soils. They are friends of the Earth and of life. We must let them be.

6

INVADERS, HITCH-HIKERS, SENTINELS, KILLERS

When we are born—and even before we are born—our bodies are exposed to a variety of bacteria and viruses and we are colonised by a number of them. By the time we enter school, we become infected or passively acquire multiple viruses, most of them, fortunately, 'benign' ones. We are also endowed with endoretroviruses that are embedded in our genome, that we pass on to our children, and they to theirs. In fact, it is these very viruses that make our gestation and birth possible. Then there are viruses that lie dormant in our cells, blood and body fluids, and only emerge when instigated. Unlike the genome-embedded endoretroviruses, these viruses turn pathogenic at some point of time in their lives (and that of their host) in order to reproduce. Infecting and causing disease, therefore, is necessary for their survival.

The most well-researched of our viral passengers is the herpesvirus family (Greek; *herpes*: to creep or crawl). Herpesviruses are large-sized viruses, with such large double-stranded DNA that the process of its integration with the host or even with bacterial genomes is very complex. We are infected by at least eight common herpesviruses, and 90 per cent of the world's population carries at least one herpesvirus in their bodies at some point in their lives. Among the most common and well-

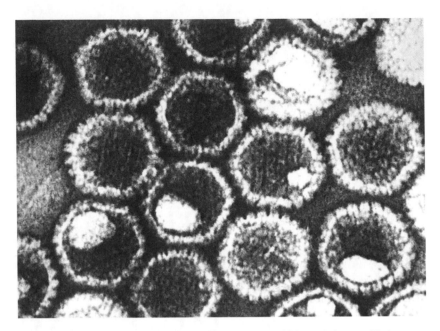

Herpesviridae is a large versatile family of DNA viruses which can infect any life form ranging from molluscs to humans. Their most common manifestation are rashes and blisters. Of the 132 known species, only nine affect humans, and despite being widely present in body fluids, most of these do not cause any disease unless they get triggered by some unknown environmental factor. Because they lay low inside us, all herpesviruses are assumed to be capable of causing latent or lytic infections, depending on the host's condition.

known herpesviruses is one that causes chickenpox (or herpes simplex virus 3 or HSV3), a disease some of us may have suffered from in our childhood. In later life, HSV3 can erupt painfully to cause shingles, and then vanish as suddenly as it appeared. Some of us in our teenage years may have had discomforting sores around our mouths that would have been caused by herpes simplex virus 1 (HSV1). When we discovered sex, a few of us may have developed rashes around our private parts, which is caused by another herpes simplex virus 2 (HSV2). Once you've got it and have taken a dose of antiviral medicine, the disease wanes and the virus retreats, but instead of total extermination, it finds refuge in some hard-to-reach places such as the nerve cells and stays put. Unlike most viruses that infect and then leave, these laggard herpesviruses, like HSV2, quell their ambition of boom-and-bust by switching off their lytic genes and staying dormant. They float around in our cells and rarely cause disease a second time. Only a shock like a high fever or a high dose of a strong medicine can shake them out of their torpor and prompt them to bring about disease in our bodies.

So how did all these diverse viruses come to live within us? There are no easy answers to that question, but we do know when and how herpesviruses got into us. The herpesvirus family belongs to a primitive lineage of DNA viruses that emerged and diverged several times. The ancestral viruses lived in fish which passed them on to some amphibians and reptiles, and they later crossed over into early birds and a few different mammalian families. They then travelled from a monkey to an ape and then to our earliest ancestors, eventually getting into us, creating a new viral species each time.

The story of the first herpesvirus crossover in humans begins around 6 million years ago. Our ape ancestors acquired the chimpanzee herpesvirus (ChHV1) which evolved slightly in each descendant as it was passed on, and eventually became HSV1 in us. Both ChHV1 and HSV1 cause oral herpes or cold sores in their respective hosts, and all monkeys and apes have their own unique HSV. We humans have two—the cold sore causing HSV1 and it's not-so-far-removed cousin, the sexually transmitted genital herpes virus, labelled HSV2.

Why do humans have a second HSV? Unlike HSV1, the evolution of HSV2 was not a straightforward hand-me-down from our evolutionary ancestors, and our present-day understanding of HSV-2 is like a scientific detective story of great finesse. Scientists reckon that somewhere between 3 and 1.6 million years ago in the savannahs of East Africa, an early ape ancestor called *Paranthropus boisei* or *P. boisei* (*para*-beside; *anthropos*-human; also labelled *Australopithecus*) fed on the carcass of a chimpanzee. This was a time when the ice caps on the Poles were spreading and Earth was beginning to cool in spurts. This, in turn, was sucking up moisture from the air and causing continental land masses to become parched. Once-moist forests were gradually turning into open scrub forests, and well-watered savannahs were becoming dry grasslands. These were difficult times for all life on Earth and it was no different for the apes and early ancestors of the hominin that were living in Africa during this time. The fickle environment made food scarce, and hominins and some apes had developed a taste for meat. The carcass that *P. boisei* devoured was probably that of a precursor of modern-day chimpanzees and bonobos. Perhaps when *P. boisei* tore open the carcass with its hands, a deep gash or a pre-existing sore on its hand enabled the virus from the carcass to infect it. Alternately, since fire had not yet been tamed, the meat of this ancestral chimpanzee was eaten raw, and the virus in the blood and flesh could have entered *P. boisei* through a wound in its mouth. Our other more immediate *Homo* ancestors also competed for food and shelter with *P. boisei* in the forests and savannahs, and perhaps they hunted and feasted on each other's flesh as well. This reinfection was passed on from *P. boisei* to one or more of our *Homo* ancestors; either *Homo habilis* ('the handyman') first, which passed it on to *Homo erectus* ('upright walking man'), or perhaps directly to *H erectus*. The virus could also have been transmitted sexually and not just from feasting on each other's flesh—a possibility that I am sure has crossed your mind. Recent research suggests that we have acquired DNA fragments from our cousins, both near and distant ones, like the Denisovans and Neanderthals. This suggests that mating between *P. boisei* and our early *Homo* ancestors may well have taken place. Irrespective of the mechanism, this crossover was much more recent than the passage of HSV1, which means that HSV2 is more closely related to ChHV1 than it is to HSV1.

Why did HSV2 become genital herpes instead of remaining as oral herpes, as was the case with HSV1? This could have had something to do with an evolutionary process known as 'niche partitioning'. Niche partitioning or niche differentiation happens when two species with identical niches or ecological roles are in competition. They remain in a constant state of war until either one or the other is driven to extinction or is pushed out by the other. When that happens, the 'loser' must adapt to find a new niche. When both species have found their separate niches within the body of a host, they are more likely to coexist and 'put down their weapons'. Since the cold sore virus, HSV1, was passed down earlier and took a liking to our mouths, HSV2 had to settle for the private regions. Clinically, the two viruses can infect both locations; they're just better suited for the ones they normally infect.

Once geneticists found the evolutionary origins of HSV1 and 2, it was assumed that the virus HSV left with our ancestors, *Homo sapiens,* during the 'Out of Africa' migration event which took place between 72 and 60,000 years ago, and that it has stayed with all human populations ever since. But this is disputed by a study published in 2020 by Italian scientists. The Italian researchers were able to precisely date the origin and dispersal of HSV1 and 2 in human populations around the world. Their analysis suggests that HSV1 migrated out of Africa a mere 5000 or so years ago, and that HSV2 reached the Americas and Europe through the transatlantic slave trade only in the eighteenth century. This study partially explains why the prevalence of HSV2 is higher in the Americas, second only to Africa, while it is not as widely prevalent in the Levant, Arabia and South Asia where early humans had settled first. However, the debate on HSV is far from over, and further research can help find answers on its origins and global spread.

Another family of viruses, the *Polyomaviridae* family (*poly-*: multiple; *-oma*: tumour) comprises at least 100 known viruses, including a dozen or so pacific residents in our bodies. Many of them are ancient viruses that opportunistically cause tumours in a wide range of hosts such as fish, birds and mammals. The JC virus (JCV) is perhaps the most common polyomavirus in humans. It was named after John Cunningham of the UK, the patient in whom it was first discovered in 1971. JCV is weakly pathogenic. It infects reluctantly, awakening only when the host's immunity dips very low or is compromised and it manifests as small bunion-like tumours, especially in the kidneys. In children with very low immunity, JCV can cause debilitating neurological problems.

We acquire JCV soon after our birth, and by adulthood most people have JCV present in their body fluids. Very few people are completely free of one or the other strain of JCV. We excrete the virus frequently, mostly through our urine. How much remains or is released from our bodies varies between populations—65 per cent of East Asian and 'Out of Asia' (or Asian-derived) populations like Native Americans shed JCV, which is more frequent than people in Africa where only 20 per cent of the population does so. Virologists can say with some degree of certainty that JCV originated in West or Central Africa and crossed over from a primate into a human possibly while hunting or consuming bushmeat, just like it happened with herpes and most recently with HIV and Ebola. It is possible that some monkeys, many primates, most apes and perhaps all the ancestors of *Homo sapiens* acquired it more recently than HSV, but what seems certain is that our ancestors already had JCV in their bodies as they were leaving Africa. The band of sapiens who left Africa had a single JCV genotype, which was passed on to all their descendants. The single strain of JCV evolved into distinct types over time in every land mass that humans settled in. Based on how humans spread and interbred, the geographic distribution of the numbers of the viral subtypes of JCV is now in excess of ~1100 worldwide strains. This diversity of strains represents important genetic 'breadcrumbs' that help archæologists and geneticists understand the migratory routes of our ancestors as well as reconcile the timing of their in- or out-migration from one region to another. Geneticists can now calibrate the time when our ancestors left Africa, reached faraway places like the Philippines and Taiwan, and colonised remote islands of Oceania. They can chart the gene flow between north-eastern Siberians and the Ainus of Japan, and the

Koryak people's contribution to circum-Arctic Americans. One of the best resolutions to the question of when humans reached North America and colonised it was done by studying skin samples of residents from Old and New World populations. We now know that the earliest human pioneers to settle in the Americas brought with them type 2A of JCV from northeast Asia when they crossed the Bering Strait between 15,100 and 13,300 years ago. After Christopher Columbus, when European settlers arrived in America, they carried primarily Type 1 and Type 4 strains of JCV, and these gradually spread among native Americans. Africans brought by the slave trade harboured Type 3 and Type 6 and these, too, have diffused into the wider population.

Another dormant, resident member of the polyomavirus family is the Merkel cell polyomavirus (MCV) which causes a rare form of skin cancer. It, too, is shed like JCV, but unlike JCV, it is spread not through urine but with the wear and tear of skin. Newborn babies acquire the MCV virus from their parents within a matter of days. This virus is so prevalent that you find it in sewage. There are other, accidentally discovered, patient-named polyomaviruses, like BK which was isolated from a Sudanese patient in 1971, and a few that are named after institutions like WU and KI (Washington University and Karolinska Institute, respectively).

In 1997, doctors discovered a new virus in patients. It took a long time to give it its current name, torque teno virus (TTV; Latin; *torque*: twisted like a necklace; *teno*: thin), which incidentally, were also the initials of a Japanese patient being treated for hepatitis. At first, it was called 'transfusion transmitted virus' but was later renamed because several different viruses are transmitted by means of transfusion. TTV belongs to the family of viruses called anelloviruses, and is widely present across populations of all ages. It is found in most of our body fluids and organs except perhaps the brain. Since its discovery, seventy-six different species of TTV have been grouped together in the family *Anelloviridae*, (Latin; *anello:* ring), some of which have been found in seals, cats and horses. When dormant, TTV causes no disease and no one really knows what it takes to shake it out of deep slumber, or even what disease it can cause. However, TTV can potentially inform doctors about the immune status of a person—the higher the numbers of TTV, the lower the immunity. TTV is a useful marker for patients who undergo a kidney transplant or other surgeries, and doctors can tell if immunosuppressive drugs are working by measuring the levels of TTV before and after an operation. By measuring a titre of TTV, doctors can also estimate how quickly a patient's immunity can be restored, post operation. Unfortunately, the use of TTV in diagnosis is not yet widely employed.

Some immunologists believe that anelloviruses, to which the TTV belongs, influence the development of the immune system and its maturation. And no matter which body fluid you extract, whether from a healthy or a sick individual, it is sure to contain this virus. One study has shown a confounding link between low anellovirus levels during pregnancy and the risk of a child developing psychotic disorders.

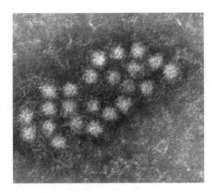

Since the discovery of the first anelloviruses in 1997, there have been at least thirty other species that have been identified that are capable of living within our bodies. Anellovirus species like TTV (left) colonise the human body within the first year of birth, after which their numbers tend to taper down.

A few viruses that infect us in childhood are capable of triggering juvenile insulin-dependent diabetes mellitus type-1 (IDDM1). Viruses like enteroviruses and rotavirus which can cause diarrhoea in infants and children, or serious childhood infections like rubella and mumps, invade and damage pancreatic beta cells, which gradually leads to diabetes type-1. And while we incriminate fizzy drinks, confection, fries and burgers for the rise in diabetes and obesity, there are at least five confirmed human adenoviruses and several animal obesogenic viruses that can induce weight gain. Adenovirus-36 (Ad36), in particular, is a human virus that causes symptoms like the common cold (which is actually caused by rhinoviruses) and spreads through the air by close contact with an infected person. Some obesity experts believe that at least one-third of overweight people are so because of Ad36. In 1994, researchers at the University of Wisconsin infected chicken and mice with Ad36 and discovered that it increased their body fat by 50 to 150 per cent, compared to uninfected animals. When they tried this on monkeys—the closest animal model to humans—all infected monkeys gained weight, despite eating and exercising just as much as a control group that was not exposed to the virus. When the study was repeated on more than 500 individuals, they found that 30 per cent of the obese individuals were infected by Ad36 and on an average gained more than 50 pounds, while 11 per cent of non-obese individuals had the virus. However, those infected with Ad36 had lower cholesterol, triglycerides and blood sugar despite the weight gain. And no, it is very unlikely that you can gain weight from a person infected by Ad36, whether obese or not. So keep all your friends close, through thick and thin. Scientists now know that once Ad36 gets into the fat cells of a person or animal, a specific gene diverts more glucose from the blood into the adipose layer. There is currently no prescribed treatment for infectobesity (here's some jargon you can use to test how updated your GP is!), but scientists hope to develop antiviral agents that can help us get rid of such viruses. Next time you are not able to shed those extra kilos, blame the virus!

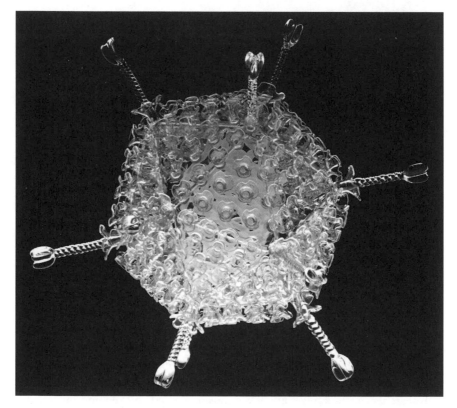

Human adenovirus 36 or Ad36 is one of fifty-two types of adenoviruses known to infect humans. First isolated in 1978 from the faeces of a girl suffering from diabetes and enteritis, Ad36 commonly causes respiratory and eye infections but its persistence in body fluids like blood can lead to obesity in animals and humans. A particular viral gene (called early region 4 open reading frame 1, or E4orf1) synthesises a protein that makes tissues accumulate fat. Several other viral infections are now considered possible risk factors for a global obesity epidemic, and perhaps require therapeutic intervention.

We know only a few hitch-hiking viruses, while a large number perhaps remain unknown to us. What use are they to us? For starters, resident viruses, especially *polyomaviruses*, make for excellent population-markers. We can now use different viruses to determine the nature, episodes and timing of intermixing in different ethnic groups. In countries with a complex racial demography such as Brazil, which was colonised in several waves by immigrants who came from far and wide, the mosaic of polyomavirus in their bloodstreams and sewage can reveal interesting stories. Because of their ability to leave behind 'fingerprints' based on ethnicity, these viruses are also being explored as tools in forensic sciences. House meets CSI.

Increasingly, immunologists are beginning to appreciate that these free-ranging viruses may be more beneficial than harmful to us. Because of their widespread yet largely dormant existence, they are an excellent indicator of our immune status and can be used to predict individual health. A few long-term immunological studies have found early evidence of some resident viruses offering mild protection and the ability to signal the possible presence of viruses and emerging medical conditions. Like spooks, some viruses like JCV spot vulnerable cell receptors that can be open to attack, alert our immune system to repair these and thus avert serious diseases. They can be used as indicators of the onset of early manageable but later irreversible conditions like cancers, diabetes and obesity. Another lurker, the hepatitis G virus, decelerates disease progression in immune-compromised patients like those with HIV. Some viruses may be key to building our innate immunity, the one that we are born with. Very few of us, apart from some isolated ethnic groups and communities, remain herpesvirus-free. Since the discovery of these silent and hidden viruses, the idea that viruses are all agents of disease is giving way to the notion that they can exert positive and even protective effects. We still don't know if there is any trade-off in those who are infected versus those who are not. And we don't quite know whether the presence of these freewheeling viruses continues to influence the evolution of our genome, shape our immunity or affect the functioning of specific cells. What we do know is that there is much we still need to find out and explore about these tiny residents in our bodies, and we need to delve deep to begin to unravel the secrets they hold.

A 7-million-year-old family portrait

From left to right 1: Homo ergaster from Kenya, 1.55 million years before present (my bp); 2 and 7: First Homo erectus from Georgia, about 1.7 my bp; 3, 6 and 13: A male Neanderthal from France, 50,000 years bp; 4, 5 and 16: A female Neanderthal with a boy from France, 40,000 years bp; 8, 15 and 11: A male and female Australopithecus from Ethiopia, 3.1 my bp; 9: Homo floreseinsis from Indonesia, 18,000 years bp; 10: Homo sapiens from Cro-Magnon, France, 30,000 bp; 12: Paranthropus boisei from Tanzania, 2.5 my bp; 14: Sahelanthropus boisei from Chad, 7 my bp; and 17: Homo habilis from Kenya, 3 my bp.

The frequent mixing among ancestors of modern humans facilitated the exchange of pathogens. Scientists are certain that before our ancestors left Africa, they had acquired the herpesvirus. HSV-1 and 2 were passed down in a pretty linear fashion, from our ape ancestor, Paranthropus to Homo habilis and H erectus to humans, and is widely prevalent in every living human. The same route was taken by viruses that cause neoplastic tumours like cancers of skin or lymph node. We know of at least thirty viral, bacterial and other microbes that have crossed over chiefly via sexual intercourse. By analysing ancient DNA, it is now possible for scientists to peek into the interbreeding between our earliest ancestors. Our ancestors who left Africa were formidable reservoirs of tropical diseases, and exposure to these new diseases may have caused the decimation of populations of naïve Neanderthals of Eurasia and Denisovans in East Asia. Scientists believe that two pathogens—Kaposi's sarcoma herpesvirus (KSHV; also known as human herpesvirus 8, which develops into a cancer from the cell linings of blood vessels) and bacterial gastric cancer caused by Helicobacter pylori may have led to the extinction of the Neanderthals. While our sapiens ancestors overwhelmingly transmitted more pathogens to their far-removed cousins, there were at least some that they got from their cousins. Human papillomavirus (in particular the aggressive HPV16 which causes cervical cancer) was passed down from Homo heidelbergensis to H neanderthalensis and the Denisovan, who passed it to our ancestral sapiens between 120 and 60,000 years ago. HPV16 has evolved to become the most prevalent sexually transmitted infection in the world. Human hair lice, the bane of schoolchildren and their mothers, too was passed down from Homo erectus to sapiens about 1.18 million years ago. Intestinal worms were passed down several times as our ape-human ancestors competed for carrion with hyaenas and other scavengers, and we acquired these wriggly creatures from carcasses of herbivores and carnivores. Tuberculosis, the most prevalent bacterial killer of adults is relatively recent in origin, and was acquired from our Neolithic ancestors between 67,000 and 47,000 years ago.

7

A SPOTTY HISTORY OF THE SPECKLED MONSTER

Our empires were mighty and proud. Their overlords commanded grand armies and amassed enormous wealth. They appeared invincible and etched their names deep in human memory. Microbes, on the other hand, are minute and insignificant, but when they found the right conditions, they rampaged through defenceless populations with incredible speed and ferocity. These creatures that no one can see have demolished cities, waylaid large armies and weakened powerful empires. Successions of epidemics have distorted the demographics of nations, leaving in their wake piles of bodies and diseased, scarred, frightened hordes of people. It sometimes took several generations before kingdoms could recover. In human history, about a dozen diseases have been able to do this to empires and cities. And one of the most devastating of them all, was the smallpox.

Smallpox belongs to a group of some eighty-odd viruses called the 'poxvirus' family. The ancestor of smallpox is believed to have originated in the Sahel—a vast tropical expanse in Africa sandwiched between the Sahara and the rainforests—which spans the continent from the Atlantic Ocean to the Red Sea. The grasslands of the Sahel are replete with ample seeds and insects and are a haven for a variety of rodents that prosper here. Among them is a timid-looking rodent called the 'gerbil', its name derived from a corruption of the Arab word *jarbu* (meaning 'anything that hops') by the French, who named it 'jerboa' and then 'gerbil'. The hind legs of a gerbil are quite long so they hop rather than scamper or scurry, making them appear somewhat like a cross between a mouse and a diminutive kangaroo. They are small and cute and it's possible sometimes to see them in pet shops. Or if you visit the dry grasslands that stretch from Africa to South Asia, you may see them hopping across the tarmac on warm evenings in the glare of your car's headlights. Gerbils are social creatures and live in communal burrows. Like many rodents, gerbils have a specially evolved immune system that protects them from the pathogens they host. They don't get fevers or infections from these pathogens but are instead protected by these microbes from common disease-causing pathogens. One particular species of gerbil in the Sahel is the naked-soled or the northern savannah gerbil (*Gerbilliscus kempi*), and it is in the lungs, blood and guts of this small furry rodent that an ancestor of the smallpox virus first emerged. Scientists believe that the pox of the gerbils (called the 'tatera poxvirus'; *Tatera* is what they call gerbils in India) was passed on to camels in which it evolved as the 'camelpox virus', and to humans in whom it evolved as two closely related strains of a virus called *Variola*, (Latin; variegated, later spotted or speckled), that cause smallpox.

Among the earliest possible reservoirs from where smallpox (left) crossed over from animals into humans is perhaps this timid mouse-like Kemp's gerbil (Gerbilliscus kempi, right) which lives in the central African grasslands. Camels evolved in the Americas and migrated over the land bridge across the Bering Strait into Eurasia around 3 million years ago. Climate change drove camels into north-east Africa rather recently, perhaps about 4000 years ago. Genetic studies confirm that the camel and gerbil encountered one another around this time, and a cowpox-like ancestral poxvirus which was capable of infecting a wide range of rodents was transmitted from the gerbil to camels from where it eventually crossed over into humans. Interestingly, in each new host species the poxvirus evolved into a distinct species. The speed of evolutionary change in poxviruses suggests that they readily find and adapt to new hosts and reservoirs, which makes it important to frequently test animals like rodents and camels for pathogens that can potentially cross over into humans.

The evolutionary leap from rodents to camels and humans occurred almost simultaneously.

For a very long time, historians and epidemiologists thought smallpox arose around 10,000 BCE in north-eastern Africa and the Fertile Crescent when agriculture first began. From there, it spread from the Levant region into China and India as trade increased between ancient civilizations. Successive epidemics of smallpox interacted with climate fluctuations and caused the decline of the Roman, Qin and Ming empires. The Arab expansion into North Africa and Spain, the Crusades, the explorations and the growing trade between Europe and Africa, India and South East Asia repeatedly reintroduced smallpox into Europe, and it was Europeans who brought the pathogen to the Americas. It is believed that King Rameses of Egypt was killed by it in 1142 BCE, as were several Roman senators and Greek and Persian kings. Princess Mary I and her husband, William II of Holland, Queen Mary II, King Louis I of Spain, several emperors and shoguns in Japan, and even Pocahontas is said to have died from it. George Washington had smallpox before he became US President. Abraham Lincoln had it while he held office and showed symptoms when he delivered the Gettysburg address. Over millennia, this devastating disease has caused political and economic upheavals and left lasting painful memories around which myths, superstition and religion were moulded.

The Spanish conquistadors unknowingly carried the smallpox virus in their lungs to the New World, decimating the natives. Large smallpox epidemics led to the collapse of the Aztec and Inca empires. In 1521, even though the Aztec army vastly outnumbered the Spaniards, it was weakened by smallpox and was massacred, which led to the fall of Ten·och·tit·lan (now Mexico City). It is said the Spaniards were exhausted butchering the weakened natives. When the Spanish invaders first arrived in 1518, Mexico had about 25 million native inhabitants; by 1620, this number had fallen precipitously to 1.6 million. 'The religions, priesthoods, and way of life built around the old Indian gods could not survive such a demonstration of the superior power of the God the Spaniards worshipped,' wrote William H. McNeill, author of *Plagues and Peoples*. 'Little wonder, then, that the Indians accepted Christianity and submitted to Spanish control so meekly. God had shown Himself on their side, and each new outbreak of infectious disease imported from Europe, and soon from Africa as well, renewed the lesson.' The history of the New World rested on just this one fateful episode. The African slave trade from the sixteenth to early nineteenth century and the import of indentured labour from India from the mid-nineteenth to early twentieth century for

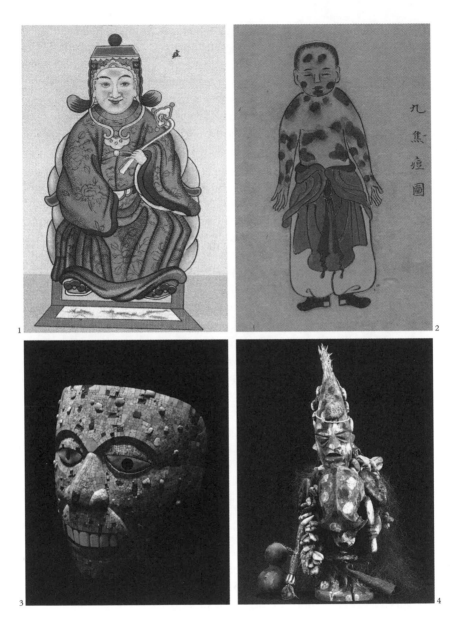

Across the globe, cultures have feared fevers like the smallpox and have prayed to placate the terrible gods who they believe are responsible for its visitations. Pan-chen (1) was the main Chinese god prayed to during smallpox epidemics, and there are temples in China dedicated to him. A textured watercolour (about c1720) from Japan depicts the pustules of the disease (2). A mid-sixteenth-century Aztec mask made of turquoise mosaic with conch shells on wood of a demi-god Xiuhtehcuhtli from Mexico shows a face disfigured by boils and pustules (3). This early-eighteenth-century statuette is of Sopona (4), a smallpox deity of the Yoruba tribe from modern-day Nigeria, Benin and Togo. Similar statuettes of Sopona were taken by slaves from West Africa to Brazil during the transatlantic slave trade.

growing sugar cane, introduced new pox strains into regions that had not previously encountered this deadly virus.

As the Spaniards and the French, and the British after them, colonised the west coast of North America from Baja California to Vancouver, the pernicious release of smallpox decimated native communities. Smallpox was the first biological weapon. In order to avoid the loss of lives in battle with native Americans, the British deliberately unleashed smallpox on them. In a letter written to Colonel Henry Bouquet in 1763, Sir Jeffrey Amherst, commander-in-chief of British forces in North America, advocated dusting scabs of smallpox on to blankets and leaving them outside native settlements that were under siege; or distributing and trading bison skins laced with the pox with native American people before the onset of winter. Interestingly, there are sixteen towns and a college in the US that commemorate General Amherst's dastardly exploits, all of which are situated in territories where native Americans once lived.

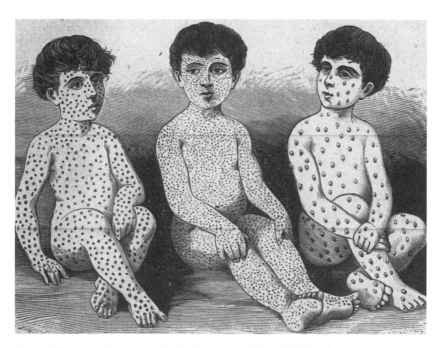

Several different pathogens manifest in similar ways. This 1880 French engraving shows three diseases with similar symptoms, often confounding physicians—all of them caused high fever accompanied by rashes and pustules. The boy on the left has measles (rougeole), the middle one scarlet fever (scarlatine), the one on the right, smallpox (petite verole volante). Measles and scarlet fever (which, incidentally is caused by a bacterium) left behind no scars after they had run their course, but smallpox would often leave its victims scarred, pock-marked and sometimes blind for life.

Until World War I, anyone who had not been inoculated (Latin; *in*: into; *oculus*: to graft or transplant a bud) had a one in three chance of contracting smallpox. Even if you survived the disease, chances were that you would be badly scarred and often blinded in one or both eyes. The scourge scythed through the population, killing youth and older folk more than children. In an epidemic, about 30 per cent of the susceptible population would contract the infection. The smallpox rash first appeared inside the jowl and pharynx of the patient, followed by small boils on the face and forearm, and by the next day the rash would erupt on the trunk and legs. This is when the patient would be most infectious, so any attendants would also acquire the infection by now. Rashes and boils would soon swell to become pustules and then burst open into painful lesions, while the fever and body ache would continue. If the patient survived the first nine days of fever and pain, the chances of recovery would be high and the pustules would soon dry into scabs. However, the new skin that replaced it was left pitted and pock-marked and if a rash or pustule erupted in or near the eye, blindness was certain. Death rates among naive populations varied between 20 per cent and 60 per cent (it was 4 to 11 per cent in populations that had encountered the virus before) and left most survivors disfigured, scarred and blind. On the plus side, it also left the majority immune against future infections.

As viruses go, smallpox is relatively large and can even be seen under a good simple microscope. Smallpox and other poxviruses have relatively large genomes. Smallpox has 200 genes compared to HIV which has ten, and Ebola only seven. Smallpox's DNA is shaped like a dumb-bell, wrapped within two membranes and overlain by rows of spikes, which makes the virus look like an overfed hedgehog. Unlike viruses like Ebola that live in gorillas, or influenza, Zika, dengue and Japanese encephalitis that have multiple intermediate hosts, humans are the *only* host that the smallpox virus can survive in. And being a large-sized DNA virus, the smallpox virus or *Variola* did not produce too many mutations per cycle. There are only two clades of smallpox (a 'clade' is a distinct group of virus species that have descended from a common ancestor; each species then creates several 'types' with minor differences in genes that are called 'strains')—the more widespread and virulent *Variola major* that killed 30–50 per cent of those who got the infection, and *Variola minor* that killed less than 1 per cent.

By the ninth or tenth century, some civilizations like China and India, and perhaps a little later, a few indigenous communities of North Africa like the Garamante and Berber people (at an undetermined time), discovered that inoculating a healthy individual with moist pus or dried skin from a smallpox pustule between the thumb and forefinger or into the arm—a process called 'variolation'—provides lifelong immunity from the disease. A person who had been variolated usually developed mild symptoms of smallpox, and once the fever had run its course, would be conferred with lifelong immunity. But the consequences of variolation were not entirely predictable and never pleasant. That it would cause fever was certain, but in about 2 per cent of the people inoculated with dried scabs of smallpox, it caused scarring and/or blindness, and even death.

1   2   3   4

Smallpox epidemics created such widespread fear and panic that the disease became embedded deep within myths, cultures and beliefs across civilizations. Throughout the Indian subcontinent there are temples devoted to fearful goddesses that people believe could cure and protect them from diseases like smallpox. In north India, the region bounded by Himalayan rivers to the north, east and west, and the river Godavari to the south, the chief smallpox deity is Śītalā mata (1). Her name literally translates as the 'cool-mother', because smallpox causes a sharp burning sensation in the pocks and her blessings are invoked to cool and soothe the victim. Her shrines were usually placed under trees where she was represented by a stone, a clay image, or a piece of wood adorned with vermilion. Some villages had more than one Śītalā mata temple with each caste or community worshipping at its own small temple. Small Śītalā mata temples were built on the periphery of villages or sometimes as part of established temple complexes. Temples were also built along caravan routes ranging from Turkmenistan in the north to Myanmar in the east, and along Hindu pilgrimage circuits in the Indian subcontinent. In all her finery, Śītalā is usually draped in a red sari and shown astride a donkey. Her emblems are a bundle of sticks or a broom (sammārjanī), a winnowing fan, and an earthen pot. Iconographically, Śītalā resembles the Buddhist demon-goddesses Hārītī (2) and Jyestha (3), who were part of the early Hindu pantheon from around the second century BCE until the tenth century CE. Names of medieval Hindu goddesses were suffixed with -ma, -mai, -mata or -amma, meaning 'mother'. The worship of Śītalā became more prominent around the early to mid-eighteenth century. South of the river Godavari, temples of goddess Mari-amma exist in nearly every major south Indian city. Emigres from Tamil Nadu who left in the mid-eighteenth century to cultivate sugar cane in Fiji, Mauritius, Malaysia, East Africa and the West Indies built temples with imposing Mari-amma statues, like this statue from the 190-year-old temple in Chinatown, Singapore (4).

Around the eleventh century, the practice of variolation was carried by traders from India and China to Persia and Turkey. The Chinese applied powdered crusts of old poxes to the nostrils. In India, Hindu Brahmins used a needle with a small inoculum of dried pox crusts to variolate. The Persians ingested dried pox crusts with wine and vinegar, and gradually, the practice of scarring the skin and introducing a few cells of dried skin or pus from a pustule became widespread. Lady Mary Wortley Montagu, wife of the British ambassador to the Ottoman Empire, who had been left pockmarked as a child, witnessed variolation in Constantinople and brought this practice to England in 1721. In England, she advocated variolation to prominent families, and following the successful inoculation of children of the royal family, many more people came forward to take the jab. George Washington introduced it among soldiers of his Continental Army in 1776.

Around 1762, a beautiful young English milkmaid in a village in Gloucestershire with a flawless complexion and a confident smile bashfully said, 'I shall never have smallpox for I have had cowpox. I shall never have an ugly pockmarked face.' A thirteen-year-old orphan boy overheard the milkmaid's boast, or so the story goes. The boy was Edward Jenner, apprentice to a country surgeon, and it is this story that was retold in an 1837 biography that has since been repeated endlessly over the years. However, it is not entirely true.

The *real* story goes like this: in the summer of 1768, a few years after Jenner purportedly heard of the milkmaid's speculative immunity, a country doctor by the name of John Fewster, like many other doctors during this time, was dabbling with variolation. He had been offered a small country home where he could conduct his experiments, and he was allowed to provide boarding and lodging to anyone willing to try his experimental inoculation regime. Among those who came forward to be experimented upon was a group of farmers from Thornbury, a village near Bristol. But when Fewster gave each of them a jab of smallpox, he noticed that the farmers did not develop the characteristic swelling or lesion that people who received the jab usually did. He realised that the farmers *already* possessed immunity to smallpox. Yet they swore they had not been variolated or ever had smallpox in the past. One farmer told Fewster that he had had cowpox recently, and upon prodding, other farmers also recalled having had some form of cowpox in their childhood or youth— but none of them had ever had smallpox. Fewster's enquiries led him to an astute and exciting clinical possibility, and although he did not record it, he shared his findings with other country doctors who met regularly at The Ship, a tavern and inn at Alveston near Thornbury.

In July 1768, Jenner was apprenticing under Daniel Ludlow, a surgeon based at Chipping Sodbury. Daniel and his brother Edward, an apothecary, were part of this close circle of country doctors. It could have been one of the Ludlow brothers who went home and spoke about Fewster's observation to their nineteen-year-old apprentice. By the time Jenner eventually became a member of the medical society that met at The Ship, he had already turned into a 'cowpox bore' for his obsession with the subject. He had begun collecting and using fluid from the lesions of mild cases of cowpox and inoculating anyone who agreed to get the jab. Gradually, he became more confident that cowpox jabs were a safer and more effective alternative to inoculating with smallpox (variolation). He spent nearly three decades thinking about cowpox and smallpox before testing his theory on 14 May 1796, by inoculating his gardener's son, eight-year-old James Phipps, with an exudate from a cowpox sore from the hand of milkmaid Sarah Nelmes. Phipps suffered a local reaction and felt feverish for a few days but then recovered fully. In July, Jenner injected Phipps with a tiny dose of smallpox to see if the mild cowpox had indeed protected him against smallpox. Phipps remained healthy.

Earlier, in 1788, Jenner had been elected a Fellow of the Royal Society for a landmark nineteen-page ornithological publication, *Observations on the Natural History of the Cuckoo*. Among several observations on British birds and their calls, nesting behaviour and diets, Jenner reported on how a newly hatched cuckoo ejects the eggs or fledglings of its foster parents from the nest. He observed that the eviction is done when a peculiar depression between the scapulae (shoulder blades) of the young cuckoo disappears about twelve days after its birth. However, when Jenner advocated with Royal Society members in 1798 to publish his discovery of vaccination in their journal, *Philosophical Transactions of the Royal Society*, he was rejected. Undaunted, Jenner decided to self-publish his discovery in a pamphlet with the rather laboured title, 'An enquiry into the causes and effects of the variolae vaccine, a disease discovered in some of the western counties of England, particularly Gloucestershire, and known by the name of the cow pox', in which he described the protective effect of cowpox against smallpox. His pamphlet was received with deep scepticism in scientific circles and faced ridicule in the tabloids and magazines. The clergy claimed it was repulsive and ungodly to inject someone with the remains of a diseased animal. Despite the initial rejection, however, reports of cowpox inoculation being tried successfully by several country physicians across the isles began appearing in local papers. Gradually, some doctors, friends and village councilmen advocated with the king and Parliament, and Jenner received a handsome grant of £10,000 in 1802 to advance his experiments.

On 17 May 1803, at the first meeting of the Royal Jennerian Society (founded by fellow doctors and admirers), Edward Jenner insisted that his friend and fellow physician, Richard Dunning, be credited for coining the term 'vaccination' (Greek; *vacca*: cow). The era of smallpox vaccination had dawned. Vaccination became a common practice in England, and gradually in Europe and the Americas too, and was introduced into British colonies like India, where it saved millions of lives.

Edward Jenner's work led to smallpox becoming the first disease for which a vaccine was developed and it laid down the essential principle upon which all vaccines rest. What is particularly interesting about the smallpox vaccine is that even before the virus that caused smallpox—before *any* virus for that matter—had been discovered or identified, a vaccine to

It took decades to convince people from all walks of life about the utility and safety of vaccination. This satirical and unflattering view of vaccination by James Gillray (titled: 'The cow-pock', or 'The wonderful effects of the new inoculation') was published in June 1802. It shows poor patients crowding into a room where Dr Jenner is giving a woman a jab. A schoolboy's coat pocket has a pamphlet: 'Benefits of the Vaccine Process'. Miniature cows sprout or leap about while a butcher protests in despair at horns that have sprung from his forehead. A worker with a pitchfork sees a cow bursting from a swelling on his arm while another cow breaks through his breeches. Others are developing carbuncles on their foreheads and chins. On the back wall hangs a painting of a yellow cow on a pedestal (reminiscent of Nicolas Poussin's painting 'The Adoration of the Golden Calf') to which a group of kneeling people pay homage. The scene combines fantasy and realism.

protect against the disease it caused was developed. It happened again when Louis Pasteur developed the rabies vaccine in1885, even before the rabies virus had been identified.

In 1967, the World Health Organization (WHO) launched a global concerted effort to end smallpox. In 1977, Ali Maow Maalin from Somalia became the last recorded natural case of smallpox. Maalin worked as a cook at a hospital and also helped as a vaccinator but he had been reluctant to get himself vaccinated, fearing the pain of the jab. After he recovered from smallpox, he advocated for communities to take up the polio vaccine by narrating his own story to people who were afraid of being vaccinated. He died in 2013 of malaria, after working for years to eradicate polio.

It took a while, but gradually, as more people benefited and smallpox epidemics declined, caricaturists became strong advocates for vaccination. In the late 1790s until 1820, satires by Thomas Rowlandson, Isaac Cruikshank and his sons George and Robert, helped in promoting wider acceptance of Jenner's vaccination campaign. In this June 1808 caricature by Isaac Cruikshank (titled: 'Edward Jenner and two colleagues seeing off three anti-vaccination opponents', the dead smallpox victims are littered at their feet) shows Jenner standing on the right between two colleagues, holding a vaccination knife whose blade is inscribed 'Milk of Human Kindness'. He addresses three old-fashioned doctors who seem to be leaving the scene disgruntled, and holding large blades dripping with blood and inscribed 'The curse of human kind'. A cherub is about to crown Jenner with a laurel wreath, saying, 'The preserver of the human race'. The ground is covered with dead or dying infants, heavily spotted. A spotted mother clasps her speckled infant, while her infected husband leans against her. On the extreme right stands a woman saying, 'Surely [sic] the disorder of the Cow is preferable to that of the Ass.'

Before the British brought vaccination to India, variolation was practised by Brahmins in
some pockets of the country which may have provided moderate protection against the disease
(1). Edward Jenner had advocated the widespread dissemination of cowpox inoculum in all

2

3

the British-Indian government impinged directly on the lives of ordinary Indians. The
British manufactured myths about the widespread prevalence and acceptance of the practice
to encourage people to adopt vaccination. Francis Whyte Ellis, a British administrator
and scholar of south Indian languages composed a Tamil verse (2), called 'The Legend
of the Cow Pox', written in the style of the puranas. It consisted of a colloquy between the
physician and the gods, Dhanwantari, and the feminine life force, Shakti, where the gods
state that a cure derived from the cow was created by them to relieve humanity of suffering
from smallpox. Ellis passed this off as an ancient text. Recent scholars have labelled this
as a 'pious fraud' that fooled many western and Indian scholars of scriptures. By linking
the vaccine to the revered cow and the gods, Ellis legitimised vaccination in the Madras
Presidency. The queen and princesses of Mysore states became champions who took the jabs
and appealed to the masses to get vaccinated (3), as shown in this 1805 portrait by the Irish
painter, Thomas Hickey. By 1901, 203 million people in British India, or about 70 per cent
of the population, are said to have been vaccinated.

The last two recorded cases of smallpox anywhere in the world were accidental and happened in the UK in 1978. Two photographers working in the anatomy department of the University of Birmingham Medical School contracted the disease, and one died from it. Upon enquiry, it was found that the anatomy department was located one floor above the department of microbiology where smallpox virus research was ongoing, and the photographer's desk was situated directly above the virus preparation area. The virus perhaps had travelled up through air ducts to infect the photographer. Following submission of the enquiry report which criticised its safety procedures and standards, the head of the department of microbiology took his own life. Incidentally, air conditioning and ducts continue to remain a major source of infection, like in the case of severe acute respiratory syndrome (SARS) which spread in a hospital in Hong Kong in 2002.

On 9 December 1979, the WHO officially declared that smallpox had been eradicated worldwide, and smallpox became the first and only human disease to be eradicated by vaccination. The only other viral disease to be eradicated is a disease of cattle called rinderpest, which is a distant cousin of the measles. Rinderpest was so prevalent that it had crossed over from cattle into wild herbivores in Africa before it was declared eradicated in May 2011. Measles, too, has the potential to be eradicated but for irresolute political will and myths fuelled by anti-vaxxers which have prevented its endgame.

This in a nutshell was the acceptable history of smallpox—that it arose around 10,000 BCE and was eradicated in 1979. Then, in 2014, Czech virologists discovered two formalin jars with four well-preserved tissues labelled 'Variola' at the National Museum in Prague. Since then scientists, along with archæologists, have salvaged traces of ancient DNA from tissues, teeth and bones and have used molecular techniques to study the genetic make-up of smallpox in order to estimate when it may have evolved, and have come to a slightly different conclusion. Let me digress to tell you how scientists use genes forensically to estimate when viruses or any living thing emerged, or how closely or distantly related they might be to one another. They do so by studying the arrangement of the units that make up genes.

All genetic information in all organisms, is stored in the master molecule, DNA. DNA is transcribed into different types of another genetic material, ribonucleic acid (RNA), which makes proteins when directed to do so by DNA. Double-stranded DNA is the genetic material for all cellular life forms, whether bacteria, archaea or eukaryotes, which includes all animals and plants. In fact, other than some viruses, no other life form has RNA as its genome. Smallpox, too, is a DNA virus. DNA is the iconic 'double helix' made up of two coiled strands that resemble a spiral staircase. Each strand of DNA and RNA is made of a chain of sugar molecules linked together with phosphate compounds (arrangements of phosphorus and oxygen atoms). The D in DNA is its sugar called deoxyribose, and in RNA, it is ribose. NA stands for nucleic acid. Each strand of DNA and RNA is made of four different substances called 'bases', attached to the deoxyribose or ribose sugar and arranged in a specific order or code, and it is this arrangement that contains all genetic information. There are four bases in a DNA strand—adenine (A), cytosine (C), guanine (G), thymine (T)—the last of which is replaced by uracil (U) in RNA viruses. A binds with T, and G binds with C through a phosphate link.

Like zeros and ones in a binary code in computing, A-T-G-C is the language of genes. These run as sequences without gaps to form genes. A careful reading of millions and billions of these alphabets and then being able to spot any variations in them (mistakes in copying or insertions) provides an understanding of when genes may have changed. All life forms inherit information in the form of DNA (except for some viruses that have RNA) that is handed down from one generation to the next. Some genes within genomes acquire small errors (called mutations) and pass them on to their descendants. Geneticists have devised rule of thumb to estimate the time frame when any such minor genetic changes occurred in any life form and its ancestors, and how their ancestors in the deep past evolved and split to form different species.

To understand this better, let us imagine the Holy Bible to be the genome, with its chapters as individual genes and the verses and psalms as base pairs (A-T, G-C). The earliest versions of the Bible from around 400 CE were handwritten by scribes and monks, and painted and gilded by artists and binders. Crouching over their desks for several hours every night and day, scribes wrote thousands of words, trying hard not to make any errors. Until the first printed version in 1455 CE, there were several versions of the Bible. Each time the Bible was transcribed, new text was added to it. Several Bibles were translated, and some were even translated back into different languages. The earliest Bible can be thought of as the 'ancestral genome'. Each subsequent Bible was transcribed in trademark fashion by each monastery. Each monastery, therefore, is akin to a 'species' and the Bibles it produces is the 'population'. The monastery's trademark text, design, cover page, art, font and binding gave its Bibles a distinct look. Minor changes would not alter the meaning of a verse or the text and might even pass off unnoticed. Subsequent copies would then carry the same error, and perhaps even add in a few more mistakes. Some of these Bibles containing the errors would then go from one monastery to another and be transcribed there, and each time they would gather more errors. If a clutch of Bible-philes (I mean lovers of Bibles here, not just books) with specialist polyglot knowledge to decipher Hebrew, Aramaic, European, proto-European, Latin and other ancient languages, and of course, English, came together and compared all the Bibles that were transcribed across monasteries over a thousand-year period, they would, through their discerning eye, be able to spot the changes that crept into the texts.

This is almost *exactly* what specialist scientists and researchers do on a greatly magnified scale when they are looking at the genome of any life form. They look for *patterns* in the arrangement of text (genes) and (mis-) spellings (mutations within the genes) and group them together. By estimating the numbers and places where errors have crept in, they arrive at a figure for mutation over time (the 'mutation rate'). And by using sophisticated statistical tools, they are able to arrive at the right sequence, enabling them to prepare lines of descent (or lineages) which tell them which one came before or after. This is how evolutionary scientists are able to determine the descent, the timing of emergence and divergence of viruses and all living creatures and their ancestors. Evolutionary biologists who study birds, for example, estimate that they evolved at an average mutation rate of approximately 1 per cent per million years, which means that any two species of birds diverged from one another at a rate of 2 per cent per million years. This '2 per cent rule' has long been regarded as a rule of thumb, and by using the 2 per cent hypothesis, it has been estimated that songbirds, for example, originated soon after the onset of the last great glacial cycles, or around 2,50,000 years ago. It is this 'rate of change' that geneticists were interested in measuring with the smallpox virus.

After the discovery of the two formalin jars in the National Museum in Prague, scientists isolated its DNA and matched it against forty-two known strains of smallpox from more recent times. They came to the conclusion that the virus in the tissue material containing smallpox had originated in the late sixteenth century. In 2016, a team comprising researchers from three continents and four universities—McMaster University, the University of Helsinki, Vilnius University and the University of Sydney—presented results of the molecular dating of a

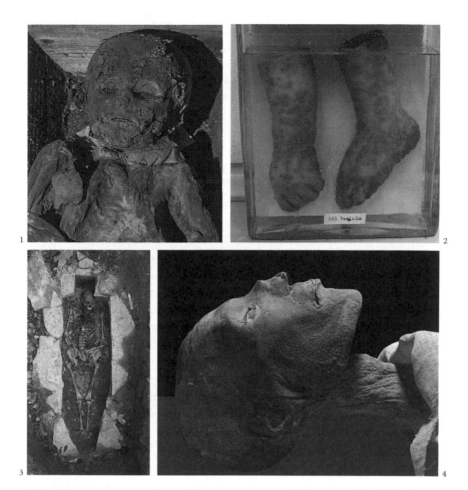

Historical accounts suggest that smallpox was present over 3500 years ago, but the further we go back in time, the less certain things appear. Recent genetic studies have helped clear up some of the uncertainties. Scientists have been able to reconstruct viral genomes from archeological samples and this has helped them find differences in the structure of smallpox viruses, as well as compare them to modern strains. The now-extinct viruses form a group, or clade, of their own. Smallpox DNA extracted from the child mummy from a crypt in Lithuania (1) estimated the date of the ancestral form of smallpox viruses to be approximately 1580 CE. Analysis of smallpox DNA from skin samples in Prague Museum (2) suggests that there was constant re-introduction of the virus from South Asia, and also by sailors and travellers who had visited West Africa and South America, which brought it back to Europe around 1650 CE. A study published in July 2020 of the DNA of poxvirus extracted from Viking-age skeletons (3) estimates the virus to date back no further than 600 CE. In light of these discoveries, it is possible that the variola virus is a relatively recent killer and not, as was thought, the reason for the death of Pharaoh Ramesses V in 1145 BCE (4). Or perhaps there were other killer strains of smallpox that we haven't yet discovered. Discovery of more well-preserved cadavers which enclose the smallpox DNA from China, India and the Levant can help improve our understanding about the evolution of the speckled monster and its impact on human history.

smallpox genome isolated from a crypt which lay buried in a church in Lithuania, of a child who had died between 1643 and 1665. By comparing it with the genetic make-up of all known modern smallpox strains, the team found that the common ancestor of the smallpox virus emerged no further back than about 1588.

This study created a stir amongst science and medical historians who had attributed many earlier historical deaths to the outbreak of smallpox epidemics, but the new findings seemed hard to refute. Then a paper was published in the journal *Science* in June 2020, which presented new evidence to show that historians may have been mostly right after all. A team of multidisciplinary scientists ranging from geneticists to computational biologists extracted DNA from the teeth and bones of 1867 humans who had lived approximately 31,000 to 150 years ago in northern Europe. Of these, only thirteen had DNA of the *Variola* virus, and analysis pushed back the time of origin of the smallpox virus to the year 603, almost 1000 years earlier than the findings from the McMaster University study of 2016 had suggested. Although the new study from June 2020 is still not able to explain the pockmarks of Ramesses who died in 1142 BCE, it does point to the fact that the smallpox virus (or *Variola*) had gradually lost its genetic diversity between the seventeenth and twentieth centuries when vaccination became more widespread. The McMaster study estimated that the two most recent strains (*Variola* major and minor) split between the mid- to late eighteenth century, which roughly coincides with the time when Edward Jenner made vaccination widespread in England and much of Europe and colonial outposts like India had taken up the practice. Even selective variolation done in societies in China, India and the Arab world would have created impediments for the virus to spread. It is clear that smallpox had slowly reached the end of its evolutionary road.

Other studies have also cast doubts on what might have caused past epidemics in ancient Egypt, Greece and Rome until early Medieval European times, and even those that occurred in the Americas. Blaming smallpox as the killer disease in *all* the instances may just have been a matter of convenience that will need to be re-examined through molecular and genetic studies. For example, when the mummified remains of a child interred in the Basilica of Saint Domenico Maggiore in Naples, Italy, around the mid-sixteenth century was examined by the team from McMaster University, they found that the facial rash earlier diagnosed as smallpox was actually caused by the hepatitis B virus (HBV). Another study is of a disease called 'cocoliztli' that swept through Mexico's highland regions in two large epidemics around 1545 and 1576, killing an estimated 7-18 million people. Several historians and epidemiologists had thought that it was caused either by measles, viral haemorrhagic fever or smallpox—all viral diseases. Genetic studies from the remains, however, have shown that these deaths were all due to typhoid, a gastrointestinal fever caused by the bacteria, *Salmonella enterica* paratyphi C. *Salmonella* is transmitted through water and food contaminated by faeces, and it is likely that the collapse of social order during the Spanish conquest may have led to a decline in civic and sanitary conditions, making the environment just right for the spread of *Salmonella*.

Genomic studies have helped improve our understanding, but the entire picture of the evolution, rise and extermination of smallpox continues to remain hazy. The only thing we know for sure is that we do not know enough. However, even partial knowledge of the convoluted evolutionary history of the smallpox virus, and an understanding of how it was blunted by variolation and finally ended by vaccination, will help us in designing more effective disease-control programmes. And while the story of the evolution and eradication of one of human history's most violent killer diseases is still a work in progress, scientists warn that there may be other poxviruses waiting in the wings. Some scientists are concerned that these have the potential to fill the yawning epidemiological gap left behind by smallpox.

In 2017, a group of scientists used horsepox genes to reconstruct the smallpox genome. They said this was neither expensive (it cost less than $1,00,000) nor difficult to do. Such reverse engineering which created a viable *Variola* virus has raised serious ethical questions and been widely criticised. As one researcher put it, the dangerous experiment does not 'justify the purported benefits', and asked for clear global norms to restrict this kind of rogue research in future. Meanwhile, there have been outbreaks of a novel cowpox from Brazil and Georgia, and a buffalopox that has recurred in South Asia where it continues to be reported as a 'fever of unknown origin'. Monkeypox, related to smallpox, originated in West Africa and the Congo Basin, but animal trade and human migration have spread it across the world. In July 2021, the most recent case of an outbreak of monkeypox emerged in China. The trade of gerbils as pets could be a possible flashpoint for the crossing over of other unknown poxviruses they may harbour. Then there is camelpox, which is found wherever naturally or introduced camels are found. Apart from these, there are at least nine species of poxviruses about which we know very little. Scientists warn that it takes just a few viral generations for a benign animal poxvirus to cross over and become a human pathogen.

On 13 December 2019, the WHO headquarters commemorated the fortieth anniversary of the eradication of smallpox under a plaque from the Global Commission for the Certification of Smallpox Eradication from 1979 which stated that smallpox had been globally eradicated. The WHO's Director General, Tedros Adhanom Ghebreyesus, said that smallpox eradication was 'a testimony to what we can achieve when all nations work together. When it comes to epidemic disease, we have a shared responsibility and a shared destiny.'

Barely three weeks later, on 31 December of the same year, WHO's country office in China forwarded a media report by the Wuhan Municipal Health Commission about a 'viral pneumonia' of unknown cause in Wuhan. Nobody knew at the time that this was the start of a pandemic of unprecedented scale. In the forty-first year of the eradication of smallpox, the words of Mr Ghebreyesus will need to be backed by all nations, as they did to end one of humanity's greatest killers. Ironically, sacred Hindu texts called *chaalisas* (or 'forty hymns') also commemorate cycles of epidemics that return every forty years, and their recitation is supposed to invoke gods and goddesses who will bestow protection against the disease. Superstition meets science meets SARS-CoV-2!

GUT
FEELING

Microbes, as we have seen, are everywhere—outside our bodies, of course, but also *inside*. And within our bodies they are found in numbers that defy the imagination.

So how many microbes are there inside us? Until quite recently it was believed that an average person harbours 10 trillion human cells that have 100 trillion bacteria within them. The origin of this mind-numbing number can be traced back to a paper published in 1972 in the *American Journal of Clinical Nutrition* by American microbiologist Thomas D. Luckey, titled 'Introduction to intestinal microecology'. Luckey's paper presented early thoughts on the evolution of the alimentary canal in higher organisms, including humans, and he hypothesised that as diets became more complex, gut microbes in various organisms created niches that became distinct for every species. He concluded that no two species, or even two *individuals* of the *same* species, have the same numbers or the same types, of microbial organisms in their guts. For humans, Luckey wrote that an average adult holds about 1000 grams of faeces in his rectum before excreting it. Then, with no supporting evidence of any kind, he proposed that there are 100 billion microbes per gram of mushy liquid in the intestine, thus arriving at the quantum of 100 trillion microbes residing in us (or rather, just in the lower end of our intestine). Luckey wrote of his admiration for 'the efficiency of these microbes in using man as a vehicle to further their own cause'.

Then, in 1977, another microbiologist—Dwayne Savage, then at the University of Illinois at Urbana—took Luckey's figure of 100 trillion and posited this with 10 trillion human cells, which he took from a textbook whose name he did not cite in his paper. Thus, a numerator seemingly plucked out of thin air put on top of a made-up denominator gave birth to an enduring microbial myth of ten bacteria for every human cell. And this number remained uncontested for nearly four decades until a 'Letter to the Editor' was published in the March 2014 edition of *Microbe* magazine, challenging this claim. The letter stated that 'the assertion of ten-fold more microbiota than human body cells was not based on hard facts'. Meanwhile, between 1977 and 2014, a couple of other developments had taken place—it had become widely accepted that microbes lived in harmony in our organs, tissues and body fluids, and together they were labelled our 'microbiota'. Another related term, but with a subtle difference, used in common parlance was 'microbiome'. When microbes and all the genes that float around in an ecosystem are accounted for together, it is called a 'microbiome'—a term made popular by eminent microbiologist Joshua Lederberg, although coined by other microbiologists before him. It is only in the past two decades that the appreciation for free-floating genes and microbes that shuffle them has grown.

Responding to the challenge posed by the letter published in *Microbe*, three scientists, Ron Sender, Shai Fuchs and Ron Milo, of the Weizmann Institute of Science in Israel, got together in 2016 to see if they could validate the 'Luckey-Savage' numbers. These scientists adopted a technique developed by physicist Enrico Fermi called the 'back-of-the-envelope method' of estimation, to help them to enumerate all the microbes in our body.

Fermi was a mercurial and charismatic Nobel prize-winning physicist who is considered one of the founding fathers of atomic energy research. He delighted in making up intractable posers that challenged his students, fellow scientists and policymakers, and then provided elegant solutions for them. The most famous Fermi conundrum was 'estimating the number of piano tuners in Chicago in the mid-1930s'. The answer to a Fermi question demonstrated how a simple process of disassembly, rather than providing an exact number, allows problem-solvers to arrive at an order of magnitude—which would be quite sufficient for the purpose for which the answer was being sought.

Using Fermi's method, Drs Sender, Fuchs and Milo set out to derive new estimates for the number of microbes in the human body. The authors obtained information on the mass and volume of tissues and cells in our body. Not everyone, they knew, has the same number of cells—the numbers and proportions of microbes vary between the sexes and are liable to change as we grow and age. They decided to estimate numbers for a 'reference' man and a 'reference' woman, a four-week-old infant and a one-year-old child, a 70-kg man, and an obese (140-kg) man. Armed with a reliable range of numbers of cells and microbes found in the bodies of each type in the group, they began their estimation. They started first with their 'reference' man, who they pegged to be between twenty and thirty years old, weighing about 70 kg (154 pounds) and being 170 cm tall (about 5'7). This 'reference man', they estimated, would contain about 24 kg of non-cellular fluids and a solid mass (organs, bones, tissue) of around 46 kg. He would hold within him 3 kg of blood, largely red blood cells which make up 84 per cent of all human cells. By contrast, massive cells like muscle and fat constitute 20 kg and 13 kg of mass, respectively. These predominate in space and volume but not in the number of cells that exist within our body. Muscle only makes for 0.001 per cent of all our cells, and fat 0.2 per cent. And these parts of the body contain very few, if any, microbes.

Using Fermi's method and data from several studies, they estimated that 'reference man' has about 39 trillion bacterial cells, give or take a few billion cells, living among 30 trillion of his human cells. This new estimate of the bacterial-to-human cell ratio at 1.3:1 is only slightly tilted towards the microbes, as against the 10:1 ratio that had been the prevalent estimate before.

The group then went about sequentially to calculate the numbers of microbes in other subjects and found that 'reference woman' had 44 trillion bacterial cells compared to 21 trillion of her own (or a bacteria to human cell ratio of 2.2:1). Next, an elderly person with the same weight as our 'reference man' was found to have a higher ratio (1.8:1) simply because the volume of blood and the number of red blood cells decline as one gets older. An obese person was estimated to have more human cells (40 trillion) and more bacteria (56 trillion) and, therefore, a slightly higher bacteria-to-human cell ratio (1.4:1) compared to 'reference man'. In terms of proportion, a four-week-old infant has more bacteria than a one-year-old infant (4.4 trillion bacterial and 1.9 trillion human cells, versus 7 trillion bacterial and 4 trillion human cells) and their ratios are comparable (2.3 and 2.2, respectively).

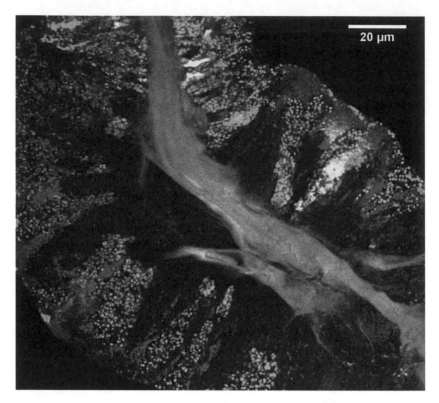

20 µm

'Stick out your tongue' is perhaps the first instruction a doctor gives when you are sick. That's because your tongue and the insides of your mouth are good indicators of infection. Until recently, microbiologists did not know how bacteria are arranged or how these affect each other. In March 2020, researchers from the Massachusetts-based Marine Biological Laboratory captured a high-resolution image of material scraped from the human tongue which mapped its bacterial population. The tongue harbours a large number and varieties of microbes that are surprisingly neatly organised in layers (called 'biofilms'). The study found that all the tongues sampled were dominated by just five genera. In this false-colour image, you can see human epithelial tissue forming the central core (pale grey), while the vivid colours on either side indicate different kinds of bacteria. The red triangular patches are colonies dominated by Actinomyces, a bacterium which rarely causes disease in humans. Forming a green crust over the surface are varied species of Streptococcus. These are common oral bacteria and many species are found in the mouth. A few species of Strep cause a sore throat and fever. The blue flecks are of a bacterium labelled Rothia, which ordinarily is harmless but causes an infection in immuno-suppressed patients. The yellow specks are of Neisseria. One species of Neisseria, but not the one shown here, causes the dreaded venereal disease, gonorrhoea. The magenta spots are Veillonella, which are common oral bacteria that rarely turn pathogenic, but when they do, they infect tissues of the heart and the nervous system. Scientists believe that these diverse bacteria on our tongue also help regulate our blood pressure! Some bacteria help ward off oral diseases especially those caused by smoking or chewing tobacco or by sustained consumption of alcohol. More research will tell us how different bacteria work together, whether bacterial communities differ across the world, and what benefits they receive from or provide to the host.

The researchers arranged the gradient of bacteria found in different parts of 'reference man' as follows: the human colon houses the most bacteria (10^{14}); followed by our dental plaque (10^{12}); next come the lower small intestine, skin and saliva (10^{11}); and then the stomach and upper small intestine (both 10^7). So if someone says that you are 'talking garbage', microbiologically, he may be right because the number of microbes at both your ends are nearly equal!

At what age do these ratios change and does this have any implications? This is an area for future research. What Sender et al. tell us is that regardless of age, gender, body type or size, we are still made up mostly of microbes. QED.

Until the 1980s, standard texts on gastroenterology and medical microbiology maintained that there were only a handful of 'good bacteria' species that live in our gut. This stable population of gut bacteria was thought to be more or less common to every human being and the predominant resident of the human gut was believed to be the bacterium *Escherichia coli or E. coli*. The first resident virus, a benign calcivirus, was discovered only as recently as 1993. Curiously, no one had delved deeply into our bodies and more importantly, into our guts to study them closely. Thanks to modern technology and methods, however, it has now become possible to rapidly scan a microgram—one-millionth of a gram—of faeces, molecule by molecule, gene by gene. Such scans reveal complete as well as pieces of genes that come from billions of microbes belonging to thousands of different species. More studies have shown us that we have been wrong about a number of things for a very long time and our assumptions based on these misguided notions may have given us incorrect ideas about how to live, how to eat and how to cure ourselves.

An important study was published in April 2011 by researchers from the European Molecular Biology Laboratory in Heidelberg. These scientists analysed the gut bacteria in humans by studying the gut microbes of thirty-nine individuals from three different continents (Europe, Asia and America), and later extended the study to an additional eighty-five people from Denmark and 154 from America. Initially, they expected to find a chaotic mixture of many different bacteria, including myriad unknown species in the human gut. What they discovered instead came as a surprise. Despite their great diversity, there was order and organization in how bacteria (and perhaps other microbes that live alongside them) were distributed in our gut. They also found that contrary to popular belief, just three genera of bacteria dominate the microbiome of our gut—*Bacteroides*,

Prevotella (once included under *Bacteroides* but recently separated), and *Ruminococcus*. Each genus can comprise tens or perhaps hundreds of bacterial species that can coexist in the human gut at any given stage of our lives. Further studies found that the proportions of these three genera of bacteria changed according to the diet of the individual. For instance, it was found that people who consume lots of protein, especially animal fats and sugars, predominantly house members from the genus *Bacteroides* in their gut. Those who consume low-protein but high-fibre diets have a greater number of *Prevotella* species, and *Ruminococcus* is predominant in the guts of those who are long-term consumers of mainly fruit and vegetables. What the studies also found was that these three types of bacteria were found in *all* human guts, even in people who suffered from chronic gastrointestinal diseases. However, the relative proportions of these bacteria changed in such individuals. The fourth bacterium *E. coli* (and others in genus *Escherichia*) which, for the longest time had been thought to be the most dominant organism in the human gut, was found to make up less than 1 per cent of the total bacterial population.

Not all species of these dominant gut bacteria are helpful. One species of *Ruminococcus* labelled *R gnavus,* if found in higher proportions, triggers Crohn's disease. The onset of disease or its acute symptoms are triggered by the release of a complex sugar by this gut microbe, which in turn stimulates inflammation in the gut lining. The Broad Institute at MIT has found the *R gnavus* phage and registered it as a possible cure for patients afflicted with *R gnavus*-induced Crohn's disease.

What this means, in essence, is that what you have eaten since you were an infant and as a child has an important bearing on your gut microbiome, which, to a large extent, is a critical determinant of your overall health.

The infant microbiome is strongly influenced by how the baby is born—whether through the birth canal or by caesarean section. Infants born through normal deliveries get colonised by organisms from the birth canal, while those born through C-section have bacteria similar to those on the mother's skin. The gut microbiome of those born by caesarean section initially lacks species from genus *Bacteroides* and their diets determine the number and type of microbial residents they will gradually acquire. Many paediatricians believe that the early infant diet is key for good health in the future.

However, food scientists and microbiologists also say that it is never too late to improve the constitution of your gut and your health by eating

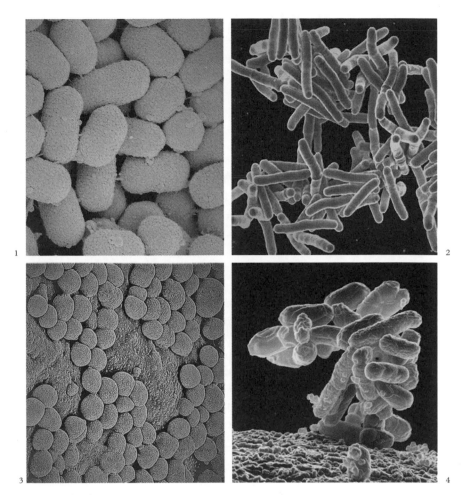

Our gut houses trillions of bacteria and other microbes. They eat what we eat, and each microbe has a specialised function. Certain microbial species like Prevotella *(1)* and Bacteroides *(2)* are adept at colonizing the mucous layer of our gut. Both these organisms are essential for the normal functioning of our gut, but if they find a way to invade the inner lining of the mouth, they can cause painful abscesses. These are dominant residents in a gut and are indicators of the food habits and lifestyle of an individual. Prevotella *is found in greater numbers in those who consume a plant-based diet, rich in fibre.* Bacteroides *are more prevalent in those who consume protein- and fat-rich diets. Another important resident is the species* Ruminococcus *(3) which helps in the breakdown and digestion of plant cell walls in the colon. Benign strains of* Escherichia coli *(4) are part of the normal microbiota of the gut, and help the host by producing vitamin K2, an essential ingredient for blood to clot. They also prevent pathogenic bacteria from colonizing our gut. Each bacterium has its specific phage which regulates their population. The various organs in our bodies are like distinct ecosystems where microbes live in harmony with other organisms. In normal conditions, these microbes work in tandem to produce a range of bioactive compounds that diffuse through our gut and fine-tune our immune system. Scientists are beginning to map the composition of the microbiome to predict health outcomes and its impact on ageing.*

'right'. Working together, gut bacteria produce around 10 per cent of the energy that the body uses. Microbiome studies among malnourished children have shown that they have fewer and less diverse microbes in their gut.

Just by looking at their gut microbes, scientists are now able to tell with 90 per cent accuracy whether a person is lean or obese. Every person leaves behind highly unique and identifiable microbial signatures (from their skin, saliva, and less so, with their gut) which could be used as a forensic tool in crime and insurance-related investigations. A housebreak in Miami was solved when one of the intruders decided to relieve himself and his microbial make-up and genes were recovered from the commode! Moral of the story: one, make sure you flush better; two, no matter what, you will leave behind a trace despite your best effort! For this reason, the human microbiome is sometimes labelled as our 'second genome'.

Our gut microflora is changing. *H. pylori* is a bacterium that lives in acidic stomachs of about half the human population. The bacterium is incriminated for causing stomach cancer and ulcers but it also maintains and regulates the gut's immune system. Until four decades ago it was widely prevalent in nearly every gut. But the indiscriminate use of antibiotics and even some preservatives in packaged foods have caused *H. pylori* to disappear and be replaced by a new set of microbes. Localised extinction of microbes within our gut can have consequences that we know little about as yet.

Bacteria make up the bulk (about 93 per cent) of all microbes of our gut microbiota in terms of mass, but other microbes also live in our guts and play a very important role in keeping our microbiota stable. In terms of mass, bacteria are followed by human-associated fungi (which thankfully are outnumbered by bacteria because we do not want opportunistic and often aggressive fungi like *Aspergillus* to take over our gut). Then there are diverse bacteriophages (bacteria-eating viruses) and other viruses that possibly outnumber the bacterial population by at least tenfold and yet make up only a very small fraction of the microbiome in terms of their weight. The rest of the mass is made up of yeasts, other microbes, as well as worms—all of which are essential to our existence. Yes, you have read it right! Worms, too, perform *essential* functions in our bodies. The saliva of roundworms, for instance, provides immune protein triggers that start a pathway which protects us from several chronic gut conditions. One such condition is called inflammatory bowel disease, which comprises two separate disorders—ulcerative colitis and Crohn's disease, both of which are debilitating and can change the course of normal life if not managed well or reversed. Tiny parasitic intestinal worms called 'helminths' that typically are absent in people living in high-income nations but which parasitise billions in low- and middle-income countries, tones down the exaggerated immunological response to antigens that trigger irritable bowel and Crohn's disease. Some doctors prescribe stool or faecal transplants as a cure. This, I realise, sounds a little off-putting, but some people actually need organisms to be seeded inside their gut flora to help them overcome chronic diseases. The few doctors that do the doo-doo transfer rely on using daily donations from healthy volunteers who have been prescreened for their gut microbes. The faeces is usually injected into a recipient several times before benefits can be seen. Although these treatments are not approved in any country yet, patients have been known to travel from one country to another in order to get themselves infected by worms that will help them overcome chronic gut ailments. The Cochrane Reviews, the gold standard of clinical evidence used to inform policymakers, says it is still inconclusive whether faecal transplant interventions should be given the seal of approval.

With so many microbes living, dividing, digesting, fermenting and dying in the gut, why doesn't our stomach or the intestine bloat and explode? Once again, this is thanks to our friend—the virus. For a long time, estimates of microbes in the human guts focused mostly on its bacteria, and any information that had been found about viruses was largely incidental. In 2003, however, a team of researchers decided to study the viruses in the human gut. They analysed all virus genotypes from the faeces of one thirty-three-year-old male. The focus of this single subject study was to estimate how many different types of viruses are released when an 'average' man does his job. The study found that nearly 1200 viral genotypes exist within the human body, and we know next to nothing about 90 per cent of them. What this means is that there are two to five times *more* viruses in our guts than there are bacteria.

This viral 'dark matter', as virologists call it, reigns over the microbiome and myriad viruses cull excess bacteria and keep populations stable. When we experience shocks to our body like a fever, through antibiotic use, by accidents and surgery, extensive travel, or when we encounter new cuisines and diets that can upset the gut's resident flora and make it unstable, it is viruses that act as mediators between microbes and our cells, tissues and organs to overcome such shocks. They help communicate between different bacterial populations and host cells which stabilises the microbiome. Viruses also help to make changes to the bacterial mix very gradually as we grow.

Due to the enormous genetic diversity of viruses, it is difficult to sequence all the viruses found in any environment, even in a microbiome such as the human gut. Although there have been significant recent advancements in genome-sequencing technologies that could make it easier, virome studies still lag way behind other microbiome studies. Until 2014, for instance, we did not know which was the most prevalent virus in our bodies. Then, thanks to the explosive growth of metagenomic databases and advanced computing, virologists discovered a bacteriophage they labelled *crAssphage* (cross-assembly phage). It is named after the computer software that discovered its genes in several publicly available databases that were used to find the viral genome. This virus is possibly the first and only organism to be named after a computer programme. *CrAssphage* regulates the most dominant bacterial species (like *Prevotella* and *Bacteriodes*) of the gut. We are yet to find the phages that regulate the other dominant bacteria, *Ruminococcus* and *Escherichia*, so clearly there is a lot more we still need to discover about our gut. There is currently no evidence that *crAssphage* causes human disease. In 2017, scientists found that *crAssphage* belongs to the largest bacteriophage family (the *Caudovirales*) and lives in a wide range of environments, like the termite gut, in groundwater, salt pans, marine sediments, plant root systems, and of course, in our own guts and faeces. Is it we humans who have contaminated these far-flung places, or does *crAssphage* naturally inhabit these disparate and unexpected places? One thing that we do know, however—which the discovery of *crAssphage* showed us—is that we know very little and understand even less about our gut microbial residents, especially the ubiquitous viral dark matter. *CrAssphage* may prove to be an even better indicator than bacteria as a marker for human faecal contamination in our water sources and food. As different strains of *crAssphage* are typed and catalogued, it can become an excellent tool for archæologists (much like other viruses such as the JCV and herpes), and may also serve as an indicator of good health and a marker for disease. The prevalence of *crAssphage* has been found to be low in traditional, hunter-gatherer populations such as the Hadza from Tanzania and Matses from Peru, compared to industrialised, urban populations. This suggests two things: first, unlike bacterial microbiota, *crAssphage* prevalence does not depend on variables such as age, sex and the body mass index of individuals. Second, *crAssphage* seems to have colonised our gut due to our industrialised lifestyle, and various strains of the phage have diverged due to population expansion. Understanding *crAssphage* will, therefore, perhaps help us understand our bodies and improve our gut health.

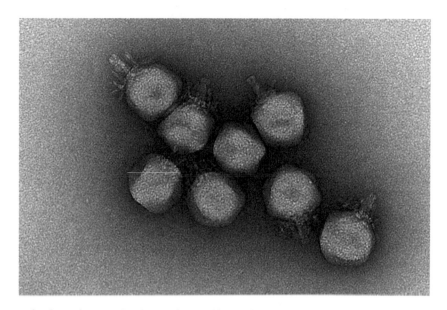

crAssphage, the most abundant and most widespread bacteriophage in the human gut, was discovered only in 2014.

The other surprise with the discovery of *crAssphage* was that its genome looked like nothing that virologists had seen before. Most of its genes and proteins from its exterior coat were not common or shared with any other virus or bacteria, and offered very few clues to the evolution of the phage.

Other than *crAssphage*, other viruses and phages in our gut are distinct within each of us. As infants we are born in near-sterile condition and our gut is rapidly colonised by microbes from our mother and the surrounding environment. At birth, the microbiomes of the mother and child are similar for a brief period but gradually begin to diverge. In early life, microbes found in faeces are relatively simple, although there can be large variation between infants. A few weeks after birth, the microbial diversity changes very rapidly with new species of microbes colonizing and becoming stable. Although we still do not know why, the first organisms that colonise an infant's gut are those which are able to tolerate oxygen in tiny concentrations, or what scientists call 'facultative anaerobes' or 'micro-aerophiles'. Gradually, these are replaced by those that live strictly in the absence of oxygen (anaerobes). The microbiome of the gut of a growing infant is affected by how it was born (whether through the birth canal or by caesarean section), its early diet, the mother's microbiota, any antibiotics that the mother may have taken during pregnancy, along with a few other factors. Viruses begin to enter the gut as soon as bacteria have claimed their space within the gut. The viral community of the guts of infants, on the other hand, contains few phages, and is dominated by only one or two types. The low diversity of the viral community correlates with low microbial diversity in infant guts. By the time a child is four years old, its microbes have become quite distinct from its mother's. Studies of twins suggest that their gut bacteria and viruses may be similar for a short while but gradually become very different from one another. In infancy, bacteriophages and 'good' viruses, in a sense, 'curate' the gradual occupancy of bacteria. As 'good' bacteria grow in the gut and other microbiomes in the body, the population of viruses, especially bacteriophages, becomes stable. As adults, we harbour within us an abundant and diverse community of viruses, dominated by DNA bacteriophages. Most RNA viruses in us are those that also infect plants. New evidence suggests that the composition of the intestinal microbiota when we were young determines whether we are predisposed as adults to allergies such as asthma.

Our microbiome is an entire ecosystem where human cells interact with trillions of microbes, and regulating these microbes are immune cells of organs themselves, and these in turn are regulated by specific bacteriophages and viruses. We still do not know what regulates viruses and bacteriophages in our bodies. Do our spongy lungs regulate the viruses we inhale? Do the bubble-wrap layers of the intestines, the stomach or bile acids or our very versatile liver, reduce the number of viruses or phages in our gut? Could there be *specific* immune cells or bacteria whose sole purpose is to keep the numbers of viruses in check? This is all a work in progress, and hopefully, scientists will find some answers soon.

As we know, all bacteria do not die by the assault of a phage; if that were to happen, it would be their death knell, not only in our gut but in *all* ecosystems. Bacteria, too, possess a microbial 'immune system' that protects them from infection by viruses and phages. This gene is called CRISPR, an acronym of 'clustered regularly interspaced short palindromic repeats'. CRISPR identifies and destroys alien virus genes. An early study estimated that nearly half of all bacteria possess some sort of CRISPR system in their genomic sequences. A 2016 paper, however, revised this estimate and its survey of 1724 microorganisms found that just 10 per cent of microorganisms have functional CRISPR. Why don't all bacteria and other microbes have CRISPR or CRISPR-like systems to protect themselves? Research suggests that the other 90 per cent adopt other forms of survival strategies—such as the ability to rapidly repair

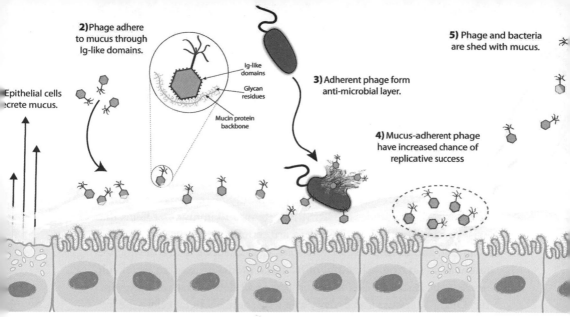

1) Epithelial cells secrete mucus.

2) Phage adhere to mucus through Ig-like domains.

Ig-like domains

Glycan residues

Mucin protein backbone

3) Adherent phage form anti-microbial layer.

4) Mucus-adherent phage have increased chance of replicative success

5) Phage and bacteria are shed with mucus.

Many of our organs like our gut and lungs are lined with mucus which helps to prevent their layers from drying out and facilitates exchange of nutrients. Most aquatic animals, too, have a layer of mucus over their skins or scales which makes them slippery and waterproof. This protein-rich mucus attracts bacteria that want to feed on it, and it is an ingredient called 'glycoproteins' that protects the mucus from bacterial attack (1). When spread out, glycoproteins carry a charge at both ends of their molecules, like a battery does. The exposed end of complex glycoproteins on fish skin or inside our gut mucus is negatively charged. Most bacterial cell walls, on the other hand, are positively charged, and bacteria therefore naturally become trapped in the glycoproteins and are digested inside the mucus before they can cause any damage and infect a tissue. The few bacteria that escape the mucus's magnetism are taken care of by—what else?—bacteriophages that are embedded in the mucus (2). Some phages directly insert themselves headfirst into bacteria and kill them. Others take a more circuitous route. They attach themselves headfirst to special defence cells called 'antibodies' (or immunoglobulin or Ig) inside the fish's mucus, with their tails protruding like Velcro. When an invading bacterium comes close, it gets an electrostatic zap from the tail, enabling the bacteriophage to counter-invade the bacterium. This action of phages has been appropriately labelled by scientists as 'BAM', which stands for 'bacteriophage-adherence-to-mucus'. The phage infects the bacterial cell, and once new phages are ready, they burst open through the bacterium, killing it instantly (3). Some of the newly emerged phages stay glued to the same fish, while others drift about until they can find a fresh layer of mucus on another fish (4). Thus, despite living in water filled with hostile microbes, fish are protected from the threat of infection thanks to the mucus that their skins secrete (5). Something very similar happens in our guts, lungs and on the surface of other organs and systems where the bacteriophages residing in our mucus work closely with our immune system to reduce bacterial colonization and protect us from infection and disease.

their genomes after viral attack. Many others live symbiotically and are protected by their partners.

Besides, CRISPR may not necessarily be the best form of defence either. A bacterium under attack sometimes acquires and activates its CRISPR sequences quite slowly and often dies before it can defend itself. Only about one in a million bacterial cells emerge unscathed from a phage infection. In a landmark 2012 paper published in the journal *Science*, the 2020 Nobel Laureates for chemistry, Professors Emmanuelle Charpentier and Jennifer Doudna, showed that CRISPR can be adapted to function in a test tube and programmed to cut specific sites in isolated DNA. This has opened up a world of wonders for the future of genetics, although we may need to proceed with utmost caution. Scientists have identified at least six distinct types of CRISPR systems and immune proteins that exist in bacteria. Each of these have arisen independently in the course of evolution and in response to persistent viral infection. Most of what we know is drawn from a handful of bacterial model systems and a lot still needs to be discovered—for example, we still don't know what triggers the CRISPR system into activity and why some bacterial cells benefit, while others don't. Or how and when bacterial cells switch on and off their CRISPR genes. Or whether or not CRISPR systems serve any other regulatory functions.

Another interesting phenomenon that several studies have shown is that there is a clear pathway linking the action of microbes and gut cells to the functioning of the brain. Physiologists call such a connection an 'axis'. The microbiome-gut-brain axis is a two-way communication system between bacteria and gut cells on the one hand, with the brain. It helps determine the patterns of our moods, emotions and, in the long term, our intelligence, from the transient to the permanent. It even helps build our attitudes and shape our perspectives. We are also discovering that the gut microbiome and brain respond to each other's instructions. It is perhaps a hidden survival mechanism that we are only now beginning to understand. The communication hotline is through the large vagus nerve, which controls the heart, lungs and digestive tract. The word 'vagus' is derived from the Latin for 'wandering', as in 'vagabond' and 'vagrant', but this nerve is anything but that.

Diet is key to maintaining a healthy microbiome and the gut-brain axis. A well-balanced diet that includes leafy vegetables, cereals, fruit, fish and lots of water is key to keeping a functional axis responsive and in an upbeat mood. The food we consume has a direct impact on the bacteria that live within us. The interplay of bacteria, phages and immune cells in the gut (and elsewhere) could provide key information on how to manage future health and disease, and how we can harness the microbiota in our pursuit for personalised medicine and treatment. With so much happening in our guts, and also lungs, vagina and skin, microorganisms

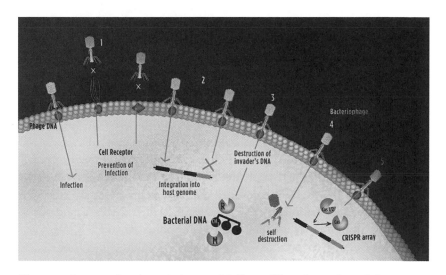

Not every phage attack on bacteria is successful. Over millions of years, bacteria have evolved several strategies to survive an assault by viruses. One simple survival tactic is by living symbiotically with another organism and getting protection from it. Scientists have identified five more defence mechanisms that bacteria have evolved in their incessant struggle against viruses. One, bacteria can thwart a virus from attaching to it by changing the receptor on its cell membrane. If the receptor is shed or its structure is changed, there can be no viral binding. Scientists have found that many bacteria develop this ability, depending on certain environmental conditions. Two, as with most other organisms, bacteria too can embed viral genes within their genomes which prevent viruses from infecting them. Such protection is termed 'superinfection'. Three is a method called 'altruistic mechanism' (or 'abortive infection systems'). Here, an infected bacterium commits hara-kiri by triggering its own rapid programmed death. This prevents the invading virus from multiplying, and the bacterium's sacrifice stops the spread of the virus to others of its kind. Four, bacteria can defend themselves against viral infection by deploying two enzymes, 'restriction enzyme' and 'methylase', in succession, which break the virus's genome strands and deactivates them. Restriction enzymes have tremendous application in molecular biology because they can excise genes from the genome of one organism and insert them into another. Five, a more recently discovered gene in bacteria called CRISPR provides an immune memory about bacteriophages they have encountered before and they incorporate this into their genome and pass it on to their progeny. This 'memory' makes it possible for bacteria to recognise a phage's DNA and to resist it when a subsequent infection comes about. In this constant arms race, there are defence and counter-defence measures deployed by the host's immune cells and bacteria, and the viruses. Many viruses 'communicate' using small protein molecules. These proteins inform viruses about their reproductive strategy, whether they should enter-replicate-burst (lytic cycle), or get embedded into the host's genome (lysogeny). It is only in the past decade that scientists have begun understanding how these protein messengers 'communicate', and this holds tremendous potential for infection and disease control.

are constantly at work, helping themselves and you, right inside you, whether you are aware of it or not. So don't reach for the antacid, that cigarette you are craving or the disinfectant you have been using compulsively. These may be counterproductive for your microbiome and therefore to you.

The answer may be to gradually change our lifestyles and not resort to quick fixes like supplements. For example, although there is a lot of research that talks of the benefits of probiotics, their long-term effects are yet to be studied. Off-the-shelf probiotics may be good when you are recovering from an episode of diarrhoea, but in general, they are medically underwhelming. Probiotic products contain proprietary strains of bacteria that have been heavily domesticated. They are present in these products in very low numbers and the species are chosen for the ease with which they can be manufactured, stored and for their shelf life, and not necessarily because they are the best for our bodies. Research is ongoing and there are no clear right or wrong answers yet. Although, as one review put it, 'It may be plausible that consuming more fermented foods improves the health of the gut microbiome and may stimulate the vagal afferents and functions of the brain.' For now, reaching for the sauerkraut, miso or kimchi may be a safer bet than that probiotic product on the supermarket shelf.

A more critical point to remember is that microorganisms, or any small life form for that matter, are not 'bugs'. Or critters, or germs. Nothing is a 'bug' when seen in the context of its role in nature and evolution. These minuscule organisms that share our bodies are not random stowaways or hitch-hikers. Some of them are critically important to our lives. We have been looking at ourselves, our existence, our biology and our relationship through a keyhole and from our perspective alone. Thankfully, new, more accessible research has thrown the door wide open to appreciate a grander view of life and processes which tie us intimately to the microbial world. Any assault on them is to turn the gun towards ourselves. The next time you follow your 'gut instinct', remember that it is not just you but a teeming multitude of organisms that may be behind it—all cheering for you! And keeping a tight control over the numbers, the spread, and the actions of all microbes is that overlord of every microbiome—the virus.

A VIRUS
VANISHES

In the Museum of London hangs a painting by Abraham Hondius with
an etched brass label: 'A Frost Fair on the Thames at Temple Stairs,
London'. This compelling, if somewhat bleak scene depicts stalls selling
countryware, food and drink, in ragged shops set in rows. The new city on
the Thames rebuilt after the Great Fire of London forms the backdrop of
the painting and the harsh winter setting is counterpoised to the charivari
of the carnival. There are entertainers and bumpkins in long grey coats
playing football and ninepins, and stilt-walkers and a fox-hunting scene
in the background. Boats on the Thames are trapped in ice and a horse
carriage trots glibly over the frozen river. A crippled beggar at the lower
right of the painting and military exercises in the background reminds the
viewer of the fleetingness of revelry in an unusual setting and in unusual
times. Hondius, a Dutch artist settled in London, painted this scene
in 1684. In a diary entry on 1 January of the same year, John Evelyn,
the chronicler of Restoration England, wrote, 'The weather continuing
intolerably severe, streets of booths were set up on the Thames, the air was
so very cold and thick, as of many years there had not been the like.'

For centuries before then, from 1250 to 1450, Western Europe had been
blessed with a mild, predictable climate with abundant crops, mostly

Abraham Hondius, *A Frost Fair on the Thames at Temple Stairs (1684)*.

well-fed populations, and few, if any, episodes of major outbreaks of disease. During these balmy times, England exported wine to France from vineyards that flourished in the southern British Isles. But gradually, in fits and starts, the climate began to cool. By 1415, the salubrious warm period had begun to wane and soon England and all of Western Europe fell under a prolonged grip of moist and frigid weather. Spikes of sharper and more abrupt cold phases began to peak around 1570 and lasted for more than 100 years. Palæoclimatologists believe that this markedly cold phase occurred due to a decrease in the Sun's activity, coupled with the most protracted stretch of intense and concurrent volcanic eruptions since our ancestors left Africa 70,000 years ago. Volcanoes eject vast quantities of sulphate-rich ash and other aerosol particles into the atmosphere, which reflect sunlight back into space. The reduced intensity of solar activity and the diminishing sunlight reaching Earth's upper atmosphere due to volcanic activity caused the whole planet to cool. This played havoc with the direction and intensity of ocean currents and caused the ice cover at the Poles to grow, cooling the Earth even more. The period shown in Hondius's painting was in the middle of the Little Ice Age (the LIA lasted from c. 1350–1850, and the years between 1450–1780 experienced the most intense cold waves), that was marked by sharp fluctuations from cool to frigid and back to cool, that caused upheavals in every corner of the world. In China, then as now the largest and most populous country, the powerful Ming dynasty fell in 1644 because of repeated crop failures and famines caused by the intense cold and weakened monsoons. In some years, skewed ocean currents disrupted sea trade. Long winters followed by short, wet summers caused hardwood trees to develop more compact pith cells and made their wood denser, which helped create the distinctive dulcet notes of Stradivarius and Guarneri violins.

These 400 years of climate change changed everything. Between 1410 and 1810, the Thames froze twenty-six times and Londoners organised 'frost fairs' over the frozen river like the one depicted in Hondius's painting. British winters were more severe than they are now and the river Thames was wider and flowed more slowly. Mean temperatures declined by 0.35 to 0.45 °C in Western Europe and England, and although this may not seem like much, it led to a long and persistent cold period with shortened summers. Even this 'minor' change was sufficient to cause crops to fail and rural economies to collapse. This dramatically changed how Europe grew and ate its food. British vineyards collapsed and the cooling forced farmers to intensively grow more winter-hardy cereals like rye, barley and wheat. In a period of just seventy years or so (from 1440 to 1510), the wine-tippling British took to quaffing whisky and gin and guzzling warm beer instead.

To feed an impoverished populace and growing cities, forests and woodlands were gradually replaced by farmland and pasture in Britain, and peasants used forest timber to trade or keep themselves warm during harsh winters. The deforestation, intensive cultivation and the storage of cereal in barns attracted rodents like rats, mice and voles, which are efficient reservoirs of a variety of diseases. The most dreaded diseases of the times were either persistent like smallpox, or more cyclical—being rat-borne and flea-transmitted—like the plague. Plague is a bacterial disease, and along with smallpox, it shaped the culture and history of medieval Europe, China, Egypt and India.

But apart from these more fearsome and better-known diseases, there was another disease that emerged in England about which very little is written or known. It caused profuse sweating, high fever, headaches, burning pains behind the sternum and the midriff, and made the sufferer extremely lethargic. Many of its symptoms were similar to the plague, but for one important distinguishing sign—those who were infected by this disease tended to have profuse, opprobrious stinking sweat, and so this disease became known as the 'English sweating sickness'.

The fever from the sickness lasted for a full day, or sometimes two, and if the patient survived the early spell, the chances of dying were lower.

The lethality among populations was particularly high at the beginning of an epidemic and the disease mostly affected robust adults. Historical records show that at least five major episodes of the sweating sickness occurred in England. It first appeared in 1483 and again in1485, when it persisted until 1492 in different parts of England. Among its victims was Prince Arthur of Wales, heir to the throne, and his death in 1502 paved the way for his younger brother who later ascended the throne as Henry VIII. The fever reached Belgium and the Netherlands through sailors in trading vessels but the infections there waned quickly, and never reached anything like the scale seen in England.

The sweating sickness struck again in the summer of 1517 when the fever appeared with a vengeance, killing nearly a third of the residents of some English villages and towns. Only children were spared. The universities of Oxford and Cambridge were closed and many scholars and students died. Aware of the disease's wrath—having seen his brother die of it—and in the midst of growing chaos, Henry VIII fled from his palace with a small retinue. Leaving his young wife behind and travelling by night, he went from manor to manor to escape the disease. Many courtiers, too, fled and some took the fever with them to the port city of Calais. But again, the disease disappeared from France after only a few weeks, and not many people died there.

A third and fourth episode followed each other quickly in 1527 and 1529, and on both occasions, the disease spread at breakneck speed along major roads and trade routes across England, but this time the intensity of the disease was much higher in north-western Europe. It travelled on ships and reached the Baltic States and Russia where some deaths were recorded, but port cities in France and Italy were once again largely spared. In Hamburg, 1100 citizens are said to have died within three weeks, and in Augsburg, 800 within six days. On the continent, the disease was labelled *Sudor Anglicus* ('Sweaty England'). Many deaths that occurred were not directly due to the disease itself but because of malnourishment and hunger, with people becoming too enfeebled to get up and eat. Each episode of the epidemic would kill between 30 and 50 per cent men and women, although characteristically, most children were spared. In England, the sweating sickness was largely restricted to southern parts and spread only as far north as Meath and Galway in Ireland. In Europe, the sickness emerged sporadically. Some historians and epidemiologists have suggested that the mysterious disease that caused lethargy and death in the trenches during the epic Second Siege of Vienna by the Ottoman

The depiction of death became a prominent genre of Christian literature, poems, dirges and rhymes and art from the Late Middle Ages (1250 to 1500) until the end of the eighteenth century. It was called 'danse macabre' or the 'Dance of Death'. The French word 'macabre' derives from the Arabic 'maqabir' or مقابر, meaning 'cemeteries'. The message that ran through all these different forms was the universality of death—it was a reminder that, whatever one's station in life, the danse macabre unites us all. Rich woodcuts and copperplate illustrations showed skeletons escorting people to their graves in a lively waltz. Such illustrations emphasised that epidemics could suddenly arise and sweep through the population, regardless of status, wealth, or accomplishments. Kings, popes and commoners alike were taken away, as in this 1502 French woodcut on death (left) which shows a clergyman and a worker being escorted away by a skeleton. These texts became popular when outbreaks and epidemics like the plague, smallpox and sweating sickness spread across Europe, and during the seemingly endless battles within England, such as the Hundred Years' War between France and England that left thousands of people hungry, diseased or dead. Macabre images and texts became a way of confronting the inevitability of death. Several images alluded to the 'sweating sickness' like this 1707 illustration (right) that shows a Dutch household with an old man who is gravely ill, lying on a bed. An alarmed physician examines his cloudy urine, a telltale sign of sweating sickness. A skeleton, a symbol of death, approaches an older woman who is sweating profusely—presumably she will be its next victim.

Empire in 1529 could well have been the sweating sickness. With one final outbreak that occurred in 1551, the fever abruptly disappeared from England.

Then, after a century and a half of dormancy, the sweating sickness suddenly reappeared in 1718, this time in Spain and France, where it was called the 'Picardy sweat', or *suette miliare* (Latin; *suette*: from sweat; *miliārius*: to have lesions of the shape and size of millet seeds) and it caused minor, localised outbreaks of varying severity. In 1881, in his *Handbuch der Historisch-Geographischen Pathologie*, the German physician, August Hirsch, tabulated 194 outbreaks of sweating sickness in France and Spain between 1718 and 1874. Official records suggest that Wolfgang Amadeus Mozart might have died of *hitziges frieselfieber* (German for 'severe miliary fever', or sweating sickness) in 1791.

In 1906, an outbreak of the Picardy sweat occurred in the Charentes region of France, affecting nearly 6000 people and wiping out entire families in a matter of days. A commission was set up to enquire into these deaths and attributed them to an infection caused by fleas that lived on field mice and suggested that *Sudor Anglicus* and *suette miliaire* were the same disease. The last recorded case of the disease occurred in a single soldier in Picardy in 1918, and then, just like that, the sweating fever vanished from Europe forever.

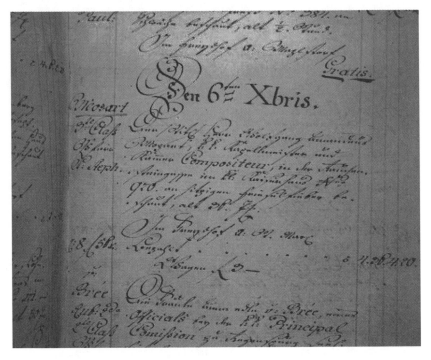

An entry in Vienna's St Stephen's Cathedral register on 6 December 1791 recording Wolfgang Amadeus Mozart's death shows a third-class funeral at the St Marx cemetery, costing a total of 8 florins and 56 kroner, with a further charge of 3 florins for a hearse. The cause of death is given as 'severe miliary fever', as sweating sickness was called in Europe. There were at least 110 deaths in St Stephen's parish from the beginning of November 1791 till the end of February 1792, almost half of which were young people under twenty-one years of age.

Apart from the suffering and mortality, one odious aspect of the succession of outbreaks was that it moulded ideas and shaped the attitudes of people towards disease, hygiene and human waste. Between the fifteenth and seventeenth centuries, a popular belief took root in England (less so in Europe) that bathing causes the pores of the skin to open up, and that this is what allows vapours to invade the body and cause fevers. The way to prevent fevers, therefore, was to lather the skin with ointments, lotions, oils, potions, astringents, and even ash, soil and dirt. For more than 300 years, most English people chose *not* to wash frequently and many chose not even to get wet! Queen Elizabeth I of England bathed just once a month; Francis Bacon devised ingenious ways to use powders and concoctions and dabbed on very little water to protect himself from the miasma; and Elizabeth's successor, James I, reportedly washed only his fingers. In some castles in England, using the moat as a cesspit proved an effective deterrent to invaders. Rivers in England accumulated so much dung and waste that they stopped flowing and stagnated. London's Fleet river collected the remains of three sewers and eleven public latrines on a bridge. Not surprisingly, the river choked, and gradually, Londoners encroached upon it and built a road, and the Fleet river became Fleet Street! Rich parts of the city of London would hire 'gongfermors' (from the Saxon word gong: go off; fermor, to cleanse) to clean latrines and waste from their areas. The waste was collected in wagons and sent outside London to fertilise farms. In 1716, Lady Mary Wortley Montagu, who first introduced smallpox inoculation to England from the Turks, remarked that the streets of Rotterdam were cleaned more diligently than her bedchambers. Avoiding water became such a mania that by the early 1600s, white linen was considered a substitute for bathing and a sign of purity and cleanliness among the upper class.

Given the seasonal variation, geographic clustering and pattern of spread of the epidemics of sweating sickness, it can be surmised with some degree of confidence that the disease was caused by a virus, and as the French commission concluded, it was most likely transmitted by mice and voles. Rats and mice are cosmopolitan and like to live wherever grain is stored. Voles, on the other hand, prefer to live in forests and riverbanks and away from human habitation. Their interaction with humans happened due to intensive deforestation. Disease experts have long tried to identify the likely pathogen and have suggested benign gastrointestinal viruses, influenza, relapsing fever and a few others that manifest similar clinical symptoms as possible causes. But the needle of suspicion points most strongly towards a virus from a family called *Hantaviridae* or the hantaviruses that cause respiratory and renal disorders, and is spread by rodents and their droppings. The name Hantan (or Hantaan) comes from a river in South Korea where the disease was first noted by the US Army Medical Corps during the Korean War. Fortunately, hantavirus outbreaks are fairly rare because you have to be pretty close to rodent droppings to become infected. Although the symptoms of sweating sickness appear to be similar to hantavirus pulmonary syndrome, the transmission of hantavirus from human to human is rare. Could it be possible that there were so many rats, mice and voles in England that it changed the dynamic of the disease? Or was the earlier strain of the virus more efficiently transmitted from human to human? Is it possible that the strain of hanta which cause the sweating sickness was more virulent five hundred years ago and gradually down-mutated to become more benign?

Despite the unlikely scenario of going down on your knees and inhaling the dust and fumes from rodent droppings, globally over 2,00,000 people a year are infected with hantavirus. Depending on the speed and quality of their treatment, 30 to 80 per cent of them die from the infection. Most cases occur in South East Asia and only sporadically in the rest of the world. Although rodents like mice have also been implicated for several diseases (bacterial diseases like salmonellosis, leptospirosis, tularaemia, plague; Q fever; and viral diseases like hepatitis, flea-borne rickettsial fever like murine typhus), none of these cause symptoms like the ones that cause the English sweating sickness, except for hanta.

In all, a total of forty-seven hantaviruses have been identified and genetic studies suggest that an ancestor of the hantaviruses diverged from another virus (from the order *Bunyavirales*, named after the town of Bunyamwera in western Uganda) and was passed down perhaps from an insect to an early mammal about 90 to 100 million years ago. Only a handful of hantaviruses are capable of causing infection in humans or even in mammals. Thankfully, not every rodent hosts the hantavirus. There are virtually no known hantaviruses harboured by rodents in Australia and very few rodents in the Ethiopian realm that host hantaviruses. But the South East and South Asia region have many new and unknown hantaviruses that can potentially jump from rodents into humans and cause a new fever. And nearly 60 per cent of all rodent species in the temperate or the Palearctic biome (which covers the whole of North and Central Europe, North America and North Asia) are known to harbour some hantavirus or the other. But we can take comfort that in Europe there have been less than ten hantavirus fever cases per year and there have been no deaths due to hantavirus in the past decade. For now, at least, the cute bank voles and field mice of Europe that have

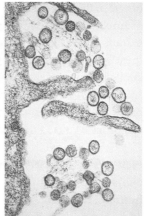

Bank vole, south England

Circular hantavirus *invading human kidney cells*

hantaviruses in their lungs, blood and guts are living peacefully away from human habitation. However, the next time a wood is logged or a pastureland is encroached, it can potentially bring mice and voles closer to humans, and a hantavirus outbreak may be in the offing. An outbreak of hantavirus (initially labelled a 'fever of unknown origin') affected the Navajo reservation community in New Mexico, USA, in 1993. Medical researchers at the Queen Astrid Military Hospital in Brussels also reported a hantavirus-like fever in a journal article in 2013.

In the story of the sweating sickness epidemics, it is fascinating to observe how a virus impacted English society and culture before it disappeared. Henry Tidy, a British epidemiologist writing in the *British Medical Journal* in 1945, concluded his observations on the sweating sickness by saying: 'Such is the disease which at one time was feared even above plague. We may bear in mind [Dr. Michael] Foster's warning: "We should be unwise to regard it as necessarily a disease nearing extinction".'

The riddle of what caused the English sweating sickness can only be resolved if traces of the pathogen's genome are discovered, perhaps in a well-preserved grave or a piece of clothing from a victim. A handful of scientists have searched for it, but are yet to succeed. It is remarkable that some viral fevers can persist for a century or so, and then vanish without a trace, only to re-emerge in unexpected places. The past is replete with such undiagnosed fevers that need to be salvaged from the dustbin of history and analysed using modern and new advanced techniques. St Vitus's dance, tarantism and St Anthony's fire occurred across Europe between the fifteenth and seventeenth centuries, causing a manic frenzy of uncoordinated dancing and shaking until people fell and even died of exhaustion. A haemorrhagic fever called *coco·lizt·li* (a Nahuatl word for 'pest' or 'pestilence') ravaged the Americas between 1519 and 1581 and caused as many deaths as smallpox did among native American people. Even today, several outbreaks and minor epidemics across the world go unreported, and then, of course, we are no wiser about what pathogen might have caused them or how we should prevent their recurrence.

New infectious diseases and fevers emerge in the background of gradual climate change, which favours their crossover from animal hosts to humans. In nature, only a small fraction of all potential pathogens evolve and become established as disease. The probability of new pathogens emerging is minuscule, but we stack the odds in their favour by our irresponsible and short-sighted actions. Not every mysterious fever emerges from dense jungles. Most new infectious diseases in humans that have emerged in the past century started from wet markets, bushmeat hunting and its consumption, from airport storage and hangars with animals in their holds, and hospitals and their laboratories where cases of newly diagnosed fevers were admitted. Economics and unregulated trade can be both a cause for the emergence of new infectious agents and a facilitator of their rapid spread. Once a pathogen crosses over, its own genetic make-up and its virulence, combined with human factors, determine how fast it will spread and how far it can reach. We have often faltered at the very first step, that of identifying an infectious agent and anticipating the risks it poses. We have been able to do even less to stop and reverse spillovers. In the age of cutting-edge microscopy, metagenomics, advanced diagnostic sciences and treatment, it is no longer acceptable that a pathogen remains unknown. An unknown fever reported by a disease control agency or a specialised lab does not count as a mystery, but a missed opportunity.

10
BEAUTY

Just as viruses have played such an important role in making us who we are, other creatures, too, have been endowed with gifts from these incredible microbes. A virus, for instance, imparted such unique qualities to a flower, it transformed it from an object of beauty to a commodity of speculation and desire, and that arguably led to the first economic bubble burst in human history!

The time was the 1630s. The Dutch Republic and the province of Holland, in particular, was in the midst of an economic boom and its cities were a vibrant entrepôt for exotic goods from every corner of the globe where humans had set foot. There was enormous wealth in that part of the world and Dutch cities attracted the finest European artists, writers and philosophers like Descartes and Spinoza who made Holland their home.

It was during this time that there arose a mania in Amsterdam and much of the Low Country for what was then an exotic flower in that part of the world—the tulip. Although tulips are now almost synonymous with Holland, they are actually native to the western slopes of the Tien Shan mountains in Kazakhstan, extending up to central Crimea. There are seventy-six known species of tulips in the wild, and in the last 200 years or so, just twenty of these wild species have been bred to yield over 3000 hybrids.

Tulips evolved from lilies about 21 million years ago somewhere in East Asia. In the wild, tulips grow in well-irrigated gravelly soils, interspersed with grasses. These perennial plants emerge from bulbs soon after winter and display flowers in a wide range of colours. Kazakh herders and nomads used tulips as food, and their beauty was first celebrated by the Persians, for whom tulips marked the beginning of spring, representing life and fertility. The earliest records of tulip cultivation are found in Persian poetry and calligraphy which described the gardens of Isfahan from around 1050 CE. Tulips became the motif for designs in various forms, from tiles to clothing, and Persian poets used the tulip as a symbol of beauty and grace. In one of his most famous verses, Omar Khayyam used the tulip as a metaphor for perfect female beauty. From Isfahan's gardens, tulips were taken to every great city of the time and were planted in the gardens of Turk and Mughal kings in Shiraz, Baghdad, Constantinople, Samarkand, Peshawar and Srinagar. By the early fifteenth century, tulips had found a place in every Turk and Persian garden. The tulip began to nudge out the more conventional flowers that were planted in beds, like asters, dahlias, carnations, hyacinths and roses. Tulip bulbs were easy to store and convenient to plant, and they bloomed in early spring and emerged simultaneously, creating uniform beds of vibrant colours. The plant in bloom had glossy dark-green leaves and large flowers which was a sight to behold for the Persians. By the late fifteenth century, formal gardens in Persia and Turkey were filled with tulips and became the leitmotif of Islam's vision of paradise.

Tulips reached Europe from Constantinople through Venice and then arrived in Bilbao, Granada and Madrid in Spain. From here, they were purchased by Dutch collectors in 1580. It is around this time that the French named the flower *tulipe*, via a Turkish word, and from earlier Persian roots of the word *dulband* or turban, because of the shape of the expanded flower. By the first half of the seventeenth century, tulips were already enormously prized by rich Dutch traders and owning high-quality tulip specimens alongside high art and 'cabinets of curiosities' from exotic lands became quite the vogue. By the 1630s, tulipmania reached fever pitch. Tulips had been growing in the Low Countries for several years, and although they had always been special, the frenzy that was witnessed in the 1630s was unprecedented and had mostly to do with some types of streaked (or 'broken') tulips that became all the rage.

The first person to document the variety of 'broken' tulips was the man who brought them to the Low Countries, Carolus Clusius, perhaps the greatest botanist and gardening enthusiast of his time. Clusius was tireless in his pursuit of exotic botanical objects and was commissioned by royal courts to seek unusual plants. During a visit to the Iberian Peninsula in the 1560s, he bought a cartload of tulip bulbs. As the favourite gardener and prospector of exotic plants for Maximilian II of Hapsburg-Austria, Clusius was commissioned in 1573 to design the royal gardens in Vienna, where tulips found a central place. Maximilian's successor, Rudolf II, however, was less enamoured of Clusius's art, believing that he had destroyed the traditional Viennese gardens. He dismissed Clusius in 1577 and set up a riding school to replace the garden. Clusius left for Holland with his cache of tulips, first creating a tulip bed in his own garden and later establishing a nursery at Leiden University in 1593. Due to his clout with Dutch and Flemish aristocrats and merchants, Clusius gradually made the tulip a plant of desire. The trove of tulips that Clusius bought from Spain spread throughout the Netherlands. In just two decades (1600–20), tulips became prized possessions in almost every home and garden in the Netherlands. They were planted in expensive Chinese and decorative Delft porcelain vases, and in Venetian crystal bowls placed on windowsills and lined in porticos and anterooms. Tulips were perhaps the first living plant to become central to the drawing room. Acquiring the most beautiful tulip became the newest fad. And among the various colours, it was the variegated kind that were most in demand and anyone who could bring a solitary specimen of beauty was richly rewarded—variegated yellows or whites on reds, dark-blue-maroon streaks on brilliant blues, yellows on brilliant blue and ghostly grey-white on white became the most sought after.

Tulips can be grown from both seed and bulb offsets. Offsets are small bulbs that grow around the base of the main bulb from which new plants grow. Flowers grown from seeds do not produce the same streaks as the original tulips, but offshoots—being clones—produced *exactly* the same striations as the flowers that the bulbs came from. Also, a plant grown from seed would take seven to twelve years to flower, but a bulb will flower the very next year. Therefore, bulbs were highly in demand. So much so that nurseries and gardens were being robbed at night of their bulbs and armed guards and patrols were deployed to stop the loot. Like art, the price of tulips was driven by whim, wealth and outright greed. A single bulb once sold for the price of a sailing ship laden with exotic goods from the East Indies. A pot with two bulbs of 'broken' tulips could fetch as much as a hundred times the salary of a skilled craftsman. A bulb of a very rare tulip could pay for five commissioned paintings by Rembrandt.

To own a rare bulb was a bit like owning a highly pedigreed stallion: valuable in its own right, but even more so because of its potential to produce identical offspring. Tulip fanciers not only collected tulips but also bought and sold the bulbs to speculators and traders. Since tulip bulbs can only be lifted from the soil when they are dormant, a futures market was created to handle contracts of purchase and sale while the bulbs were still in the ground. The business and its supply chains were based on trust, and a large rural market for bulbs was created through a network of speculators and traders. But as demand grew, so did the illicit trade, counterfeits and outright fakes of bulbs.

The market would warm up in mid-February of each year when the bulbs would sprout and informants would feed the grapevine about their prospects to merchants and middlemen. Speculators and middlemen

Wild tulips in the Tien Shan mountain range, Kazakhstan.

waited for updates from informers who patrolled gardens and small farms to get news of the crop. Some bulbs changed hands up to ten times in the course of a single day! Haarlem, a major textile trading city in the Netherlands, became a throbbing centre for the tulip bulb trade. Speculators, buyers, middlemen and even middle-class gentry from Amsterdam, Leiden, The Hague and Antwerp hired agents to sell them highly pedigreed bulbs from Haarlem. It all came to a head by the end of the winter of 1636, when thousands of people within the united Dutch provinces, including cobblers, carpenters, bricklayers and woodcutters, joined the frenzied trading, which took place in smoky taverns and back rooms. As winter ended and the first signs of tulip shoots bursting through the soil appeared in the spring of 1637, whispers of a bountiful crop of rare, streaked tulips began to reach the markets and rumours of a bumper crop spread like wildfire.

1   2   3

Speculation became rife. Traders predicted that monotones would dominate the market and that there would be fewer streaked flowers and advised their clients that the scarcity of streaked tulips would fetch them a very high price. But the speculators were proven wrong. By early March, single- *and* double-coloured tulips began to flood the markets, at first from farms around Haarlem, and then from other parts of the Low Country. With each passing day, prices began to spiral down. Almost overnight, the tavern trade disappeared. Many tulipomaniacs, middlemen and moneylenders realised that the madness had lasted too long. As the demand for broken tulips dissipated, panic sales escalated, prices plummeted even further and speculators defaulted on their promissory notes. The market for tulips collapsed on its own intricacies and frailties. By early April of 1637, the bubble had burst, leaving widespread bankruptcy in its wake. A number of small traders, businessmen and burghers in Amsterdam and cloth merchants and bulb growers in Haarlem went bust.

The streaked tulip crisis created a minor blip in the era termed 'the Golden Age of the Dutch economy'. For economic historians, tulipmania (also labelled as tulipomania) became a case study of the first 'economic bubble' in global trade. The frenetic speculative trade in tulips was more about money than culture. It was surprising, given the prevailing Dutch

From the late sixteenth till the end of the seventeenth century, northern European still-life paintings and particularly those by Dutch masters, were among the most sought after by art collectors. Within this genre of still life, paintings of bouquets and floral arrangements were a prized subgenre. They provided an opportunity for artists to show off their skills, especially their ability to capture how light fell on textured objects, and to flaunt their knowledge of a wide range of floral specimens. Almost without exception, early bouquet paintings were both highly realistic as well as improbable, because many of the flowers depicted together could not have been in bloom at the same time. Fresh colours contrasting with dark, funereal backgrounds gave these paintings considerable decorative appeal. Before the late 1580s, tulips were rarely shown in still-life paintings. And even when they were featured, they were usually with other flowers in a bouquet, like in this 1524 painting by Jan Brueghel the Elder (1). However, in the decades that followed (about 1595-1637), as the resplendent tulip became central to the Dutch economy, it also became central to still life, as seen in this 1618 painting by Christoffel van den Berghe (2). Just before and during Tulipmania, there were paintings that featured only tulips, reflecting the feverish speculative market in tulip bulbs and flowers in pots. At the same time, there was an undercurrent of Calvinist frugality within Dutch society and art, and some painters had a more sober and moralistic view of tulips, as in this 1603 vanitas by Jacques de Gheyn II (3). The skull, large mirror that hangs like a bubble and hourglass convey the transience of life and the certainty of death, and remind the viewer of the futility of beauty, pleasure and wealth. The name of this art form is derived from the opening verse of Ecclesiastes in the Latin Bible, 'Vanitas vanitatum, omnia vanitas': vanity of vanities, all is vanity. Vanitas persisted as an art form even after tulipmania ended as a stark reminder of society's folly of greed and excesses, as shown by this 1671 work by French artist Philippe de Champaigne (4). After tulipmania, tulips gradually lost their prime position in bouquets, garlands, wreaths and cartouches and were pushed aside or replaced by more fashionable specimens. Many masters painted fewer, and sometimes no tulips in their bouquets, like this 1827 painting by Cornelis van Spaendonck (5), reminiscent of Brueghel's bouquet made 300 years earlier.

belief in strict Calvinist values, that an entire nation was driven by an obsession to bet their life's savings on a streaked flower.

So what was it that caused some tulips to become 'broken' (streaked), while others remained plain-coloured, or as they would be called in modern design and fashion parlance, 'colour-blocked'? Botanists and tulip bulb collectors discovered that 'broken' tulips were ephemeral, and over three or four generations, would lose their stripes and return to monotones. Tulip breeders realised that the fashionable mottling of the petals was not a result of selective breeding or some special cultivation technique but an environmental variable that was beyond their understanding and control. The unexplained phenomenon of tulip 'breaking' also reduced the lifespan of the bulb, thereby increasing its rarity and pushing prices even higher.

Early British gardening literature from the 1910s distinguishes three types of tulip flowers of this kind—'full break', in which colour is removed in areas of the petals except in the yellow and white ones, 'self break' in which the colour is intensified, and 'average break' when 'self break' appears in a fully coloured petal. Although by now botanists and gardeners had begun to suspect that it was some kind of pathological invasion of the bulb that was causing the streaking, experiments had failed to find either a fungus or a bacterium in 'broken' bulbs. The possibility that tulip 'breaking' was caused by a virus was proposed only in 1928 when Dorothy Cayley, a fungus expert at the John Innes Centre in Merton, UK, found that when a streaked tulip bulb was grafted with the bulb of a monotone tulip, a quarter of the resulting tulip flower also became streaked. She also induced tulips to become streaked by inserting a plug of pounded-up tissue from a streaked tulip bulb into a monotone tulip bulb. Cayley suggested that this was because of a 'virus or enzyme infection'. In nature, this infection was likely to be spread by aphids, those small fleshy green, brown and black mite-like sap-sucking insects on plants. One specific aphid, the peach potato aphid (*Myzus persica*), was widely prevalent on fruit trees like peaches which were planted in large numbers in Dutch and British gardens. These trees acted as reservoirs for aphids and the infection that was readily transmitted to tulips.

Later studies found that it was a virus called 'potyvirus' that caused the streaking in tulips. Since then, 214 species of potyviruses have been discovered, of which nine cause a 'break' in tulips (two of them are more serious than others). There are also several unrelated viruses that infect lilies and tulips that can cause 'breaking', though not as intensely as

This 1640 painting by Jan Brueghel the Younger titled, 'Satire on the Tulipomania' shows buyers, planters and speculators as monkeys dealing in tulips. Monkeys negotiate, monkeys weigh bulbs, monkeys count money and monkeys do the administration. The monkey on the left tallies his list of rare tulips, his sword denoting his upper-class status. On the right, a monkey urinates on tulips. A lavish business dinner is under way in the middle. At the rear, a monkey sits uncomfortably on a horse like a nobleman. In the mid-foreground, another monkey with an owl on his shoulder—symbolizing foolishness—draws up a bill of sale. In the denouement at the right, fellow speculators in debt weep in the dock. A frustrated buyer brandishes his fists, while at the back a speculator is carried to his grave. This painting was made a decade after the collapse of the market, and not only lampooned speculators as brainless and conceited monkeys, but served as a reminder to viewers of the folly of speculating on transient and fickle blooms.

the potyviruses. Recent genetic studies estimate that the potyvirus first appeared 15,000–30,000 years ago when it crossed over from a Eurasian grass host into lilies and tulips. The virus found new hosts during the birth of agriculture about 9500 years ago in the Fertile Crescent, and spread with the spread of agriculture to other regions over the next 2000 years. How 'breaking' happens in the petals and leaves of tulips was understood only as recently as the late 1980s.

When aphids infect a plant with the potyvirus, some viral genes begin the replicating and reproductive cycle. A few genes attack and suppress the production of a dominant pigment called 'anthocyanin' in the tulip's petals but leave the base colour unchanged. Each viral infection creates a unique and different type of streaks, stripes, feathers and flames because

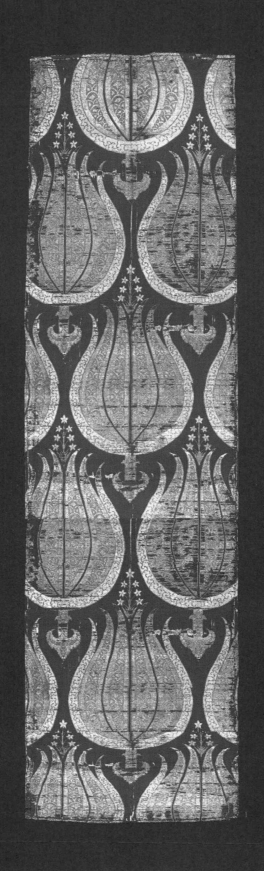

of the irregular distribution of anthocyanin in every petal. No two infected tulips or even petals have similar patterns; they are as unpredictable as the personalities and emotions of people. And as unique too.

The desire to produce 'broken' tulips petered out by the beginning of the nineteenth century. Streaked tulips are no longer planted in gardens and fields because they are considered reservoirs of disease for horticultural and agricultural crops and for other tulips as well. The Dutch revived the cultivation of tulips early in the twentieth century and built a global

2
3

A love for tulips re-emerged in the late mid-sixteenth century in Persia and Turkey, although some argue that it had never really ceased. The form of the perfect tulip in Persian and Ottoman cultures differed from the Dutch idea of beauty. For the Orient, the most beautiful tulip was one with a long, thin stem, and monotone dagger-shaped petals, not the streaked tulips that became the rage in Holland. Like the period dubbed 'Tulipmania' in the Netherlands, the flower's name characterised a historical period, the Lâle Devri or Tulip era (1718–30) in the Ottoman empire, and the artistic ferment around the tulip endures to this day. Ottoman textiles (1), ceramic tiles, books, carvings, furniture and jewellery featured stylised tulips. Calligraphers transposed the letters of the Turko-Persian world lâle (tulip) which resembled Allah and the crescent (hilal) on this delicate glass vessel used in a mosque that has tulips drawn in red (2). The tulip dominated over the rose, carnation and hyacinth in Turkish gardens and gave birth to a motif called saaz (Persian for 'music'). Rare and beautiful tulips and bulbs were used to curry favour with the Sultan, in the hope of attaining high rank in the court (3). Commoners, however, needed to pay a tax to own a tulip. In 1730, the resentment against the perverse tulip economics of the Sultan led to a bloody rebellion that displaced the overindulgent Sultan and eventually, to the restoration of economic sanity.

trade for bulbs, and the tulip became associated more with Holland than its native Persia (Iran), Turkey or the low hills of Kazakhstan. Since the revival of the global market, the tulip market is dominated by monotone Darwin tulips, and breeders have to comply with strict sanitary measures to prevent 'outbreaks'. Some kinds of streaked or striped tulips that are sold in markets today are caused by genetic variation, not by a viral infection. For those who yearn to get their hands on a true virus-streaked tulip, there are stable variants like Rem's Sensation, a flower with a deep red base that gently blends into a pure white petal. To many, such tulips don't match the spellbinding beauty of the streaked tulip in its heyday, but the virus has shown how the flower can be transformed. It is now left to tulip fanciers and breeders to recreate the magic.

Tulips are not the only flowers that are transformed by potyviruses into Cinderella. When infected with the potyvirus, camellia flowers grow more

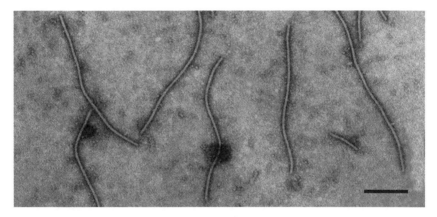

Slender, wire-like potyviruses (above) manipulate pigments called 'anthocyanins' and cause variation of colours either by diffusing or concentrating the pigment in the outermost waxy layers of the petal. Flamed, irregularly striped or streaked flowers were the most sought after and were grouped into three aristocratic pedigrees depending on the dominant hue—the rosen or rose (reds and pinks on a white background); violetten (lilacs and purples on white) and bizarden (red or violet on a yellow background). In this acclaimed still-life painting by Ambrosius Bosschaert the Elder made in 1615 (left), you see all three major pedigrees and their cultivars. At the very top of the royal rarities was the most expensive tulip ever, the rosen Semper Augustus, with its red flames on white (second from left). A tall, slender stem carries this large flower well clear of its leaves and shows off its vivid colours to best effect. Those fortunate enough to actually see a specimen of Semper Augustus in bloom found it a living wonder, as seductive as Aphrodite. The closest rivals to Semper Augustus included the Viceroy, a big, bold, purple-flamed flower (second from right). Gradually the Dutch started to adopt more patriotic and religious labels for their flowers, or commemorated the names of gardeners who created new varieties such as the Admiral Pottebacker (left) and the bizarden tulip, Root en Gheel van Leyde or the Red and Yellow of Leiden (right).

petals, develop beautiful whorled patterns, and within a few generations, also develop variegated colours in their petals. Virus-infected camellias, roses, lilies, violas, salvias, narcissi and other daffodils, canna, gladioli and many other flowers become more attractive to pollinators, and their pollen is dispersed more widely. Some viruses induce 'bunching' in flowers, and a few make some flowers bigger and fragrant. The resplendent ivy-leaf geranium develops pink-and-white streaks when it is infected by a virus of another flower, the pelargonium flower break virus. Viruses infect these flowering plants every few generations and endow them with the power to make dazzling colours without harming the plant in any way. Most of these viruses are naturally transmitted by aphids and moths. So next time you see aphids on a rose bush or camellia, please don't reach for the bug spray immediately, because they may well be in the process of making something extraordinary!

It is not just flowers but fruits, buds and leaves that also often benefit from a virus's benevolence. Some viruses that promote flowering in their natural hosts also trigger early and gregarious flowering and fruiting in grapevines. Farmers in Japan use a benign virus of apples and pears to speed up the breeding of grapes for desired traits and production. And while we are still on the serious matter of grapes and wine, here's another interesting fact about viruses—for centuries, two major viruses of the grapevine have persisted because most vineyards in the world have been created using only a handful of vine cultivars. Australian plant virologists have found that they can increase berry weight and volume by using a mild strain of one of these viruses called 'grapevine leafroll (GLR) virus'. In Japan, mild GLR strains help reduce acid content in Merlot grapevines. In France, mild strains not only protect Cabernet Sauvignon grapes but also help lower their acid yield and sharpness. In Italy, they help Dolcetto grapes to produce wines with a soft bouquet, higher plum aroma and a rich violet colour. In the future, careful deployment of select viruses may help improve the quality and yield of produce, avert the outbreak of other more serious pests and therefore reduce pesticide use. There may be new fruit notes in your next glass of wine, thanks to a virus. Let's raise a toast to that.

HOW A VIRUS SAVED A GIANT

Before 1900, or so the saying goes, a squirrel could hop from north-eastern Maine to south-western Georgia across an unbroken canopy of chestnut trees, without ever touching the ground. This could be a bit of an exaggeration, but it roughly describes the natural range—an area the size of the Netherlands—of a resplendent forest once dominated by the majestic American chestnut tree. The American chestnut *Castanea dentata* (Greek; *Castanea*: nut of Castan, a region within Asia Minor; *dentata*: toothed, because of its zig-zaggy leaf margin) was prized for its timber used in making log cabins and for its sweet fruit that was a customary holiday nibble at Christmas. Chestnuts belong to a genus (*Castanea*) from the same family as beeches and oaks (*Fagaceae*) that grows in cool northern temperate regions and comprises nine species, each of which as a full-grown tree is grand and stately. The American chestnut is a beautiful tree at any time of the year—viridian green in summer, Titian's famous yellow in the fall, and a regal leafless silhouette against bleak winter skies. A mature American chestnut rises straight up and branch-free for nearly 50 feet, can grow to 100 or even 150 feet tall and is capable of attaining a trunk diameter wider than a sedan car. It grows in acidic soil and towers over maples, oak, beech and hickory, with an understorey of blueberries and azaleas. Chestnut wood is light both in colour and weight, fairly hard, strong and stiff, yet easy to hew and shape.

In the late nineteenth century, as immigrants flooded into America, they often brought with them trees from their own countries. Botanical gardens, too, brought in exotic plants to showcase them, while plant prospectors and nursery owners sold them commercially. Several exotic species of chestnut arrived in the US from France, England and other parts of Europe, China, Japan and India. There is no actual record of this, but somewhere between 1885 and 1904, a consignment of seeds of the Japanese chestnut (*Castanea crenata*; 'with blunt teeth') arrived on the east coast of America bringing with it the chestnut blight fungus—

This 1918 photograph of a grand American chestnut tree (1) in Vermont was perhaps the last surviving tree of its kind in the state. By 1921, forests across the eastern United States resembled this ghost forest scene from Virginia (2), note a car at the bottom right for scale). The immediate advice given by state forest departments to forest-dependent communities to stop the spread of the disease was to cut down every tree and destroy all nursery seeds and sapling stock in affected areas. In less than fifty years (from 1904 until 1950), a huge forest area the size of Holland was left without a single grand tree (3). Incredibly, the American chestnut still survives in its former range, but only as sprouts from the old stump and root systems. The roots are resistant but as the tree grows, it succumbs to the fungus.

Cryphonectria parasitica. Cryphonectria (Greek; similar to *crypto*: hidden; *nectria*: sweetness) is a sac fungus belonging to a family that includes yeasts, morels, truffles and penicillin. This hardy fungus has an insidious and persistent way of parasitizing a tree. It bides its time in harsh winters, hiding under the bark or inside a lesion (ecologists call this 'overwintering') where it stays like a dormant mat of brown threads (the mycelium). When spring arrives, the mycelia awaken and produce tiny, curly, yellow-orange tendrils that bear grey-brown spores. Even a gentle wind can liberate the spores from the turgid match-head-like fruiting bodies which are by now bustling with spores. Once free, the spores waft like smoke from an idling cigarette carried by the wind until some land on an open wound or a crack in the bark of a tree. Every tree is prone to minor injuries—from an insect probing through the bark for a rich vein of sweet sap, a woodpecker hammering into the trunk for beetle larvae, a squirrel nibbling on the bark, a gust of wind snapping a twig, or a sudden freezing—any of which can open a wound and cause the sap to ooze out. This ooze becomes a substrate for the spores to germinate upon. A single spore produces threadlike hyphae—the fungal equivalent of roots—that extract nutrients from the sap and begin to sprout long filamentous cells that, at first, grow laterally to cover the open surface of the wound and gradually drill their way through the dead bark and outer layer of the trunk to reach the soft underlying tissue (the cambium) where they begin to proliferate. The fungal strands continue to penetrate deeper and deeper until they reach the plumbing system of the tree where they hijack and take over the water and minerals meant for the large growing mass on the exterior of the tree. Within one season, several small orange scabs appear over the bark. Each of these compete and eventually cooperate to make a single colony. Over the next couple of growing seasons, the fungal mass festers and bloats up the bark. The brilliant orange blight growing at the base of the tree swells and begins to look a bit like the foot of a person suffering from elephantiasis. If there are several sites where the fungus has inserted its hyphal roots, the tree develops multiple bunions that can join together to become one giant bunion. A tree girdled around its breadth by a single large canker or by several smaller ones is slowly constricted by them and starves. Whatever the age of a tree, once the disease establishes itself, the part above the site of infection begins to age rapidly and its leaves start wilting. The grand tree appears to droop, a typical symptom of the blight, hence the name. Chestnut blight is among the three most devastating arboreal diseases of temperate forests, along with Dutch elm disease and white pine blister rust, all of which are caused by fungi.

The outbreak of the chestnut blight in the US was first noticed at the Bronx Zoo in New York City in 1906 when two grand old chestnut trees that stood outside its main entrance began to show signs of the illness. Soon the entire avenue and then parks around the Bronx reported the withering and dying of chestnut trees. The blight spread rapidly in the expanding suburbs of north-eastern America. A small grove of giant chestnuts on the crest of Dumbarton Oaks in Washington DC was obliterated. The ridge on the hills north and west of Atlanta became barren. It was as if an alien hand was picking off the trees one by one. Nothing that the foresters did seemed to be able to stop or even slow down the spread of the blight. Some foresters estimated that the disease spread at a rate of about 40 to 50 miles per year. Among the people who lamented the sickness and demise of these majestic trees was the great twentieth-century American poet and farmer, Robert Frost, who chronicled the sad fate of the trees in his 1936 poem 'Evil Tendencies Cancel':

> *Will the blight end the chestnut?*
> *The farmers rather guess not*
> *It keeps smouldering at the roots*
> *And sending up new shoots*
> *Till another parasite*
> *Shall come to end the blight.*

Frost's forest-scape was dying a slow death and the gaunt silhouettes of mature chestnut trees were beginning to look like ghosts. By 1911, the spectre of chestnut trees becoming extinct seemed to be a dire possibility, and foresters, large landowners and naturalists were seriously concerned. The governments of Connecticut, Vermont, Virginia, New York and Michigan issued directives to farmers and estate owners to bring down any infected tree. The American Boy Scouts were tasked to scour forests to look out for sick trees and for signs of the blight growing on the bark of trees. Farmers were asked to burn down seriously infected or dying trees. Pennsylvania even tried in vain to create a quarantine zone in the western half of the state. But the efforts of the government, universities and the public were too ineffectual to save the forests and the tree. By the 1930s, forests appeared like an apparition of their former glory, and by 1940, of an estimated 4 billion trees, over 3.5 billion mature trees had already been lost to the blight.

The chestnut blight was the single greatest ecological crisis in North America since the great extinction of megafauna mammals—the mammoths, mastodons, giant beavers among others—that had occurred between 12,000 and 9,000 years ago and was caused due to overkill by human settlers and changing climate. When Nat King Cole sang 'Chestnuts Roasting on an Open Fire' (officially called 'The Christmas Song') in 1945, there were very few native chestnut trees left in America. Chestnuts that Americans eat now are imported from Portugal and China.

The effects of the near demise of the American chestnut from the eastern corridor forests were devastating. Chestnut's bur-like fruit were a major source of food for animals. Its decline first reduced the population of squirrels and chipmunks, and gradually, turkeys, bobwhites and cottontail rabbits, too, began to disappear from the forest floor. The absence of chestnut flowers and bark led to the extinction of at least five native species of moths and gradually made several species of insects like cicadas and butterflies locally extinct. The disappearance of chestnut leaves, bark and nuts slowed the recovery of the native white-tailed deer population which was already facing a crisis due to the expansion of agriculture, roads, railway, and cities, as well as from game hunting. Their low numbers, in turn, impacted the populations of cougars and bobcats that preyed upon them. With the dominant trees dead and gone, the numbers of the many birds that depended on them for nesting, like Cooper's hawk, woodpeckers and the cerulean warbler, also declined. Bear and racoon populations thinned down. It is said that the numbers of local hogs that fed on chestnuts from the forest floor (and made for a prized sweet and smoky ham) also began to dwindle. Tannin, a plant compound found in abundance in chestnut trees which was used for tanning leather, also declined, and by 1940, was replaced by chemical substitutes. Several native American folklore and cultural practices linked to chestnuts disappeared. The absence of large trees also impacted the flow and quality of surface and groundwater, and this affected aquatic life in ponds and rivers. Leaf litter and carbon capture declined significantly as well. Before their demise, American chestnuts dominated the hardwood forests of the eastern US. At least a fourth of all trees in the forest were chestnuts. However, a recent estimate found that chestnuts make up less than 5 per cent of all trees in their native forests of the US, and a 2013 paper labelled American chestnuts of the eastern hardwood forests 'functionally extinct'.

Interestingly, the fungus *Cryphonectria* that originated in Asia and created such havoc for the American chestnut, caused a much milder form of the blight in chestnut species imported from Asia. Although no studies about the reason for this were done at the time, it appeared that chestnut species from China and Japan had co-evolved with the fungus, their genes had coexisted for a very long time, and therefore although these chestnut trees were not completely immune to the fungus, they had developed tolerance to it. In an attempt to save the American chestnut tree, breeders tried to cross American chestnut trees with Chinese and Japanese species, hoping that the protective genes would appear in their progeny, enabling them to withstand the fiendish fungus. In the 1930s, a project by the US Department of Agriculture (USDA) attempted different combinations of hybridization. The hybrids grew quickly at first, surviving the assault of the blight, and then, after twenty to thirty years when the trees were only about 50 to 60 feet tall, they stopped growing. It then dawned upon tree breeders that Chinese and Japanese chestnuts tend to be shorter, and in acquiring the genes for resistance, some of the desired characteristics of the American chestnut—their majesty, crown size, the density of their canopy and the tree's grand architecture—had been left behind. The hybrid trees also allowed competitors to dominate them and did not develop a clear erect bole. Their fruit, too, lacked the full-bodied flavour of American chestnuts and had an acrid aftertaste. Disappointed with these results, the USDA terminated its chestnut cross-breeding programme in 1960 and the tree was consigned to its fate.

In time, the blight crossed the Atlantic into Europe but it happened later, and its effect was much milder there. It was first recorded in June 1934 around the hills of Genoa, and in 1938 the blight smothered chestnut forests in Georgia (then part of the USSR). Over the next decade or so, the fungus expanded its reach, affecting forests around Poitiers in south-western France in 1946, Switzerland in 1951, Greece in 1963 and the UK by 1974. Compared to the US, however, fewer mature and old trees died in Europe. Chestnut trees attacked by the fungus developed only superficial cankers (a kind of tumour in woody plants and trees that looks like severely swollen joints) that did not penetrate the heartwood. Mature trees nearly always recovered completely, and juvenile and young trees, too, were not severely impacted. By the early 1950s, some Italian foresters reported that chestnut trees were beginning to ward off fungal infections, although foresters continued to protect unaffected and young trees by painting their bases with gypsum and lime, a practice that produced mixed results. The impact of the blight was markedly less in Europe, but still caused appreciable distress. In the late 1960s and early 1970s, the disease forced the migration of several communities in north-eastern Greece who could no longer survive without its timber or nuts, and by 1980, the production of chestnuts in Greece had declined by 40 per cent.

Why was the effect of the fungus not as devastating in Europe as it was in the US? In evolutionary terms, the European sweet chestnut (*C. sativa*) lies intermediate to Asian and American species. The European sweet chestnut had been exposed to the fungus earlier from its Asian cousins, and it was surmised that over a few centuries it gradually developed partial immunity to the fungus. However, the story of the European chestnut's immunity was deeper and more interesting than that, and it had an unlikely hero as its central character.

In 1949, an Italian plant pathologist, Antonio Biraghi at the University of Florence's Forestry Institute, was tasked to survey the damage caused by the fungus. He reported in 1951 that some European chestnut trees showed 'spontaneous healing' of cankers. But the report went largely unnoticed even in Italy. Undaunted, Biraghi continued to collect evidence of the mortality and early signs of recovery from other countries and persisted in advocating concerted action across Europe. News about Biraghi's work reached French mycologist, Jean Grente, at the National Institute for Agricultural Research (INRA) in Clemont-Ferrand. In 1952, Grente met Biraghi in Genoa and together they explored the forests around Naples and northern parts of Italy, collecting samples from the bark and stumps of diseased and recovered chestnut trees. Grente took these back with him to his laboratory and discovered that there were two separate strains of fungus. The fungus isolated from seriously infected and dying trees was the typical vibrant orange. He labelled this strain V for 'virulent'. The second strain was extracted from the spongy tissue of recovered trees, and when introduced in a wound of a young chestnut tree, it grew very slowly and formed a pasty grey-white patch. He marked this strain HV for 'hypovirulent' (*hypo*: less). Grente was not sure what could have caused the fungus to morph into a pale version of itself, but in 1956, encouraged by his finding, he and his colleague, S. Berthelay-Sauret, smeared the less virulent HV strain on the virulent orange strain in chestnut trees in forests around Poitiers over a period of three years. To their immense excitement, they discovered that this reversed the damage to affected trees and prevented their death. They reported their findings in 1961, and several countries in Europe began to use an HV strain of the fungus to tackle the chestnut blight, producing promising results.

For many years there was not much understanding of the difference between HV and V strains of the chestnut blight fungus (*Cryphonectria parasitica*). Then in 1977 it was discovered that the HV strain had— what else? —a virus embedded in its cytoplasm! The virus was a double-stranded RNA virus and was categorised as a 'mycovirus' (*myco*: fungus). Could the presence of this virus adequately account for the difference in virulence between the two strains? This question, too, remained unanswered for the next fifteen years, and it was only in 1992 that the role of the virus became apparent. Scientists discovered that it was the hypovirulence of the virus that caused changes in the fungus rather than any resistance that emerged from the tree itself. The virus was labelled *Cryphonectria hypovirus-1* (CHV1). It is transmitted naturally or through application only to closely related fungal strains. By repeatedly pasting the benign grey fungus on the virulent orange one, chestnut trees were gradually saved from the blight damage. The hypovirulent strain not only arrested the spread of the virulent one, but also converted virulent colonies into benign ones by infecting them. However, because the virus is present in the cytoplasm and not in the nucleus of the fungus, all spores of successive generations do not necessarily become hypo (or less) virulent. Therefore, it takes a while for the virus to infect the fungus and it does so one tree at a time—and not an entire forest in one go.

Moreover, although *Cryphonectria hypovirus* (CHV) is found naturally in chestnut forests *all* over the world and is often transmitted within the spores of the fungus *Cryphonectria*, not all strains of CHV are able to provide protection against the fungus. The CHV strains found in the US, for example, have virtually no impact on protecting against the blight. The fungus itself has more than 200 strains and only CHV1 is able to infect all strains. CHV1 is also not prevalent in the UK, northern France or eastern Georgia, and has had to be brought in as a preventive biocontrol measure against the blight. By applying CHV1 to American chestnuts in the Connecticut Agricultural Station's greenhouse, very gradually, over two decades or so, they found that blight-affected cankers in trees subsided and that hypovirulence is slowly spreading. Genetic engineers are now looking for new ways to speed up the transfer of hypovirulence in America's eastern woodlands.

The principle of hypovirulence can be a viable strategy to fight against other deadly fungal plant diseases too, especially in trees. Therefore, mycologists and virologists are teaming up to prospect for other viruses that can stymie dreaded fungal epidemics like Dutch elm disease, white pine blister rust and oak blight—all of which are capable of causing immense devastation.

And chestnuts, too, may not be quite out of the proverbial woods just yet. Another Asian invader, the Oriental chestnut gall wasp (*Dryocosmus kuriphilus*), has entered Europe and North America with immense potential to play havoc. The wasp lays its eggs in fleshy shoots and leaves

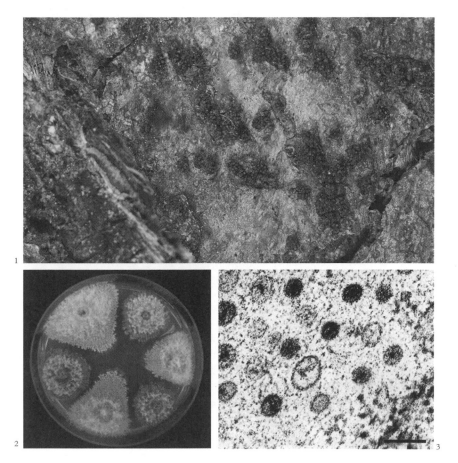

The bright orange virulent version of Cryphonectria parasitica *growing on the bark of a young chestnut tree (1). When strains are collected from diseased trees and cultured on a petri dish in a lab, the different hues of orange and grey become visible (2) which reveals their virulence. The spherical CHV-1 which confers hypovirulence can be seen inside the cytoplasm of the fungus (3).*

of the tree and its larvae feast on it. What is perhaps of even graver concern is that the wasp is a prolific carrier of the blight fungus. But can this characteristic of the wasp be used to any advantage? Can the wasp be made into an unwitting superspreader of the hypovirus-laden fungus? It is early days yet, but this is precisely what scientists are hoping to achieve.

In a nutshell, a virus may prove to be the saviour of some of America's majestic chestnut trees and it has shown us that hypovirulence can play an important part of any strategy to revive giants like the chestnut. Perhaps the lives and vitality of several other keystone species and the ecosystems they influence also hinge on viruses, microbes and smaller creatures. And the lessons learnt from the chestnut blight can help reduce the time to find effective and targeted viral strategies.

As we begin to rewild our forests, some trees will need more than just human agency. As in the case of the chestnut, it may well be viruses that will restore the giants of our forests to their former glory.

12

ZOMBIES

On 9 August 2014, seven days before the summer Olympics were due to begin in Beijing, China's otherwise stoic official news agency, Xinhua, issued a rather strange press release. There had been talk in political corridors of a possible threat of Ebola reaching China from Africa, and the article focused on dispelling the myth that drinking a mixture of coffee and raw onions will cure an Ebola virus infection. Interestingly, the article digressed into what would happen if someone did get infected by Ebola. According to the press release, a person who had died from Ebola could, 'a few hours or even days later suddenly come to life and enter an extremely violent state, tearing at and biting anything that moves, including people and animals'. Don't worry, the article said with utter seriousness, most people suffering from Ebola lose a great deal of blood and are thereby greatly weakened, and therefore their aggression is unlikely to be very damaging. The zombie like thing, the piece concluded, 'can only happen in movies'.

Not surprisingly, this article created a flutter, and although after a series of corrections the story was set right, hacks on China's rumour-prone Internet gleefully fed on the confusion surrounding Ebola-induced 'zombie disease', which, incidentally, is how Baike, China's version of Wikipedia, described the disease!

The story was far too exciting to end there. In October 2012, a leading English newspaper in Liberia, *New Dawn Liberia,* reported that 'two Ebola patients who died of the virus in separate communities in Nimba County have reportedly come back to life. The victims, both females, believed to be in their 60s and 40s, had died of the Ebola virus in Hope Village Community and the Catholic Community respectively, in Ganta, Nimba.' This is the first incident of dead victims resurrecting, reported the paper, with a flourish worthy of the best zombie novels. The website of the paper and later some papers in the UK too, featured the story, and zombie enthusiasts shared photographs and memes to celebrate it and this incredible hoax went 'viral'.

Zombie theories have flourished outside famous Hollywood capers much before 2014. In 2012, the *Telegraph* asked the UK's Ministry of Defence, in all seriousness, about the level of the government's readiness in the event of a 'zombie outbreak and attack', to which the ministry replied: 'Any plans to rebuild and return England to its pre-attack glory would be led by the Cabinet Office, and thus any pre-planning activity would also take place there.' In 2009 and 2010, the US Strategic Command conducted officers training exercises for developing scenarios of zombie attacks (called CONPLAN 8888) which lists using hand sanitisers and UV rays to combat such an eventuality, and the need for the army to target different kinds of zombielike flesh-eaters with 'firepower to the head'. Perhaps not surprisingly, the US Centers for Disease Control and Prevention (CDC) has, since October 2016, fielded a dedicated page on how you can survive a zombie threat. In September 2016, a twenty-six-year-old resident of Mumbai asked the Government of India in all earnestness about the 'readiness of our government in the event of invasion by aliens, zombies and extra-dimensional beings'. 'Can we do without Will Smith?' was the follow-up question. India's Ministry of Home Affairs failed to see the irony or the humour in the question and tersely called it a 'hypothetical situation' for which it had no 'specific information'.

While invoking zombies and preparing for a zombie attack may be taking things a little too far, we humans actually do have a long history of having our brains manipulated by parasitic organisms. The most common mind-altering parasite—*Toxoplasma gondii*—comes from cat poop. *T. gondii* is a protozoan and a distant cousin of amoeba and the malarial parasite. Its complex life cycle starts when two *T. gondii* fall in love and want to hit it off. The most congenial place to canoodle and produce their young ones is in the intestinal lining of a cat, and of course, their progeny are released with the cat's poop. The babies are encased in ball-like cysts that help them survive outside the comfort of the gut. Their hope is to find their way into another cat's gut and they do this by getting stuck to the cat's whiskers or mouth when the cat is sniffling around grass or when it is eating. Sometimes, in order to reach the cat's intestinal lining, the babies need to take a longer route. They are ingested by a pigeon or, more often, a mouse. Once inside the mouse, a few babies enter its bloodstream to reach the brain, where they alter the mouse's behaviour to make it reckless and foolhardy. So instead of avoiding places that smell of a cat or even feline urine, as an uninfected mouse would normally do, the infected mouse becomes completely oblivious to the presence of cats. The microbes manipulate the mouse's brain cells to make it feel no fear—a phenomenon that animal behavioural scientists call 'manipulation hypothesis'—a bit like Jerry in the eponymous cartoon. Unfortunately, in this case, with a much less favourable outcome for the mouse. The reckless mouse gets eaten by a cat and *T. gondii* finds its way into the cat's intestine. *T. gondii*'s ability to manipulate the mouse's behaviour, therefore, helps it to complete its life cycle.

Two Toxoplasma gondii *parasites residing in shells they created within a human cell. Their mating game is about to begin.*

Where there are cats, there are usually people. And a staggering number of approximately 2 billion people globally are infected by *T. gondii*. It is estimated that about 6 per cent of Norwegians to nearly 60 per cent of Brazilians carry the organism in their gut. While they do not display any obvious symptoms, men and women infected with *T. gondii* have been associated with a greater probability of meeting with car accidents and suffering from mental illness, neuroticism, drug abuse and suicide—which probably means that *T. gondii's* mind-altering properties also affect humans to some extent. On the other hand, one positive effect of the infection, according to a study published in the Proceedings of the Royal Society, is that those infected with *T. gondii* possess a stronger entrepreneurial drive!

In expectant mothers, a *T. gondii* infection can cause spontaneous abortion and miscarriage. Interestingly, the human foetus is only susceptible to attack by very few bacteria (like rubella), protozoa (*T. gondii*) and viruses (like Zika). Otherwise, the foetus is largely protected from disease because of the placenta which acts as a physical and immunological barrier, preventing pathogens from reaching it. The most common misconception about the placenta is that it is a part of maternal tissue. It is not. It is genetically matched to the foetus, whether male or female. In fact, maternal blood and foetal blood never cross into one another. The placenta forms within a week of conception and matures from a sheath initially made up of a single layer of cells surrounding the fertilised embryo into a much more complex layer of tissue. Organisms that infect a foetus can be detected through a single test called TORCH-z (*Toxoplasma gondii*, *Rubella*, *C*ytomegalovirus, *H*erpesviruses, *Z*ika virus) in the first twelve weeks of pregnancy. In the course of pregnancy, the mother's antibodies begin to cross into the placenta to protect the foetus. This activity intensifies closer to the end of pregnancy, and just before birth, the mother sends most of her immune cells to her child, keeping very little for herself. This is one reason why expectant mothers are more susceptible to malaria in late pregnancy.

There are some other brain-meddling pathogens that can make us behave a bit like zombies. In 1951, a disease was reported from the remote highlands of New Guinea which captured the imagination of pathologists and the public the world over. It came from an isolated tribe called the South Fore (pronounced Foray) who, for centuries, had practised cannibalism as a ritual. The entire village or extended family would cook and eat a deceased elder. These mortuary feasts—where men consumed the muscles and innards, and women and children ate the brains—were an expression of respect for lost loved ones. But over generations, this custom of eating the brains of their own kind wreaked havoc on the Fore people. In women and children particularly, it caused a debilitating degenerative illness of the brain and nerves. Locals call it *kuru*, which means 'trembling with fear or fever' in native Fore Guinean. Early surveys of Fore villages found that kuru killed approximately 25 per cent of the female population in the South Fore. And most women who survived, whether young or old, needed sticks to support themselves because severe damage to their nerves caused them to lose control of their muscles. In some cases, kuru-affected men and women who were otherwise lethargic would suddenly become aggressive before going back to sleep, a condition also occasionally seen in extreme cases of rabies. (Despite stories that abound, this behaviour has never been seen in Ebola patients). The impact of kuru was significant on the population of these gentle people. In some villages, very few women of marriageable age survived, leaving many orphans and men without wives. In some cases, kuru had a long incubation period, and it could be over fifty years after infection before the first symptoms of the disease were manifested. And often kuru's victims were infertile. With few females and not too many fecund men, the Fore people faced a serious population bottleneck and were staring at extinction. Fortunately, by the mid-1950s, cannibalism was prohibited by the government and by 1957, transmission of the infection ceased. In 2007, the last infected person died of the disease and since then the Fore people have remained free of kuru disease.

Although kuru began to wane in the '50s, the underlying cause of the disease was not discovered until 1976 when molecular research found that eating a dead person's brain activated a deadly protein that affected the brain. Since Fore women and children ate the brains of deceased people, it explained the much higher prevalence of kuru in them, as compared to the men. The protein that became infective was called 'prion' (from a word coined in 1982, short for 'proteinaceous infectious particle'). Prion disease came into focus in the early 1990s with the emergence of prion disease of cows (which was also transmitted to those who ate infected beef) called 'mad cow disease'.

Mad cow-diseased cows become infected by eating brain material from adulterated cattle feed, and humans got the disease by eating prion-infected beef sausages and hamburgers. Prion in humans leads to the rapid onset of a degenerative neurological disorder called Creutzfeldt-Jakob disease (CJD, or the 'human' mad cow disease), characterised by a loss of muscle control, personality changes and memory loss leading to rapid dementia, and finally, to catatonia. Once prions have perforated the brain,

· *This 1951 photograph of a Fore child afflicted by kuru (above) who is unable to walk or sit upright without assistance, and is severely malnourished, is among the last photographs of a victim of the disease.*

it gives it a sponge-like appearance (hence the other name for the disease is 'spongiform encephalopathy'), just as in cows affected with mad cow disease and scrapie in sheep. No antibiotics, steroids, chemotherapy or immunotherapy are effective against prions.

Scientists from University College London have been at the forefront of research on prion disease since the emergence of mad cow disease. In June 2015, they published a paper on the Fore people and prions. They found that as the Fore people gave up the practice of cannibalism, they gained a minor yet irreversible genetic resistance that protects them against any prion-causing fatal brain disease, including kuru and mad cow disease. Prions are naturally manufactured in the bodies of all mammals. However, when there is an infection, these prions fold and deform in a way that makes them turn against the bodies in which they live and then begin to behave just like a pathogen attacking the cells. The misfolded protein or prion is capable of infecting proteins that surround

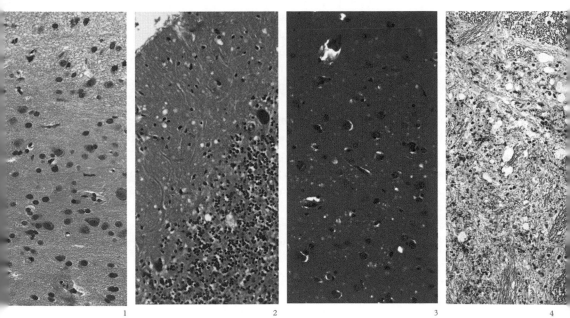

When normal 'grey' and 'white' matter of the human brain is stained with dyes (1) it shows a few tiny specks of white spaces. In contrast, a cross-section of a brain of the kuru-affected person (2) shows several small but widely dispersed white spaces, a telltale sign of a prion infection. Human brain tissue appears 'spongy', with dead brain cells surrounded by white empty spaces which is typically seen in a case of Creutzfeldt–Jakob disease (CJD or the human mad cow disease [3]). Notice that similar large, white dead spaces are seen in the brains of bovines afflicted with mad cow disease (4).

it, reshaping them to mirror its structure, and thereby, taking over the nervous system. But when the UCL team diligently scanned for genes that encode the prion-manufacturing proteins, they found something among the Fore people they had never seen before. Where vertebrate animals and all other humans have an amino acid called 'glycine', the resistant Fore had a rare amino acid called 'valine'. Could this minute alteration in one gene be enough to prevent prion-producing proteins from being manufactured, thus protecting these individuals from prion diseases and perhaps even from dementia? When the team of researchers tested this by imparting the tiny genetic mutation in mice, they found that the mice gained protection from the two diseases, kuru and CJD (or mad cow disease). When they further genetically modified mice to produce only the variant (valine) protein, they found that all mice acquired resistance to *every* prion strain (eighteen in all) for which they had been tested.

The Fore aren't the only people to demonstrate prion resistance. Australian public health scientist Michael Alpers, who had studied kuru at the Papua

New Guinea Institute of Medical Research since the 1960s, and who conducted similar research on prion protein genes in humans worldwide, found that people in far-removed communities in Europe and Japan also had the same genetic protection, indicating that cannibalism was once much more widespread and that its lasting imprint has been left behind in our genes.

Kuru is an elegant example of Darwinian evolution at work where immunity against an infectious disease has evolved within ten or twelve generations—merely a few ticks in the grand evolutionary clock. The study of kuru has helped in our understanding of how mad cow disease is transmitted and has led to the end of the abominable practice of feeding cattle with minced and dried cattle entrails. The epidemic of mad cow disease triggered an interest in degenerative diseases of the brain and nerves, and the remarkable discoveries about genetics and protein-prion that have since been made may one day help us find a cure for Alzheimer's, Parkinson's and similar diseases.

Zombies—also known as walkers, Zed, Zs, biters, geeks, stiffs, roamers, Zeke, ghouls, rotters, Zoms, and runners, to enthusiasts—have become important characters in our cultural landscape. The idea of zombies originated at least a few centuries ago, and the earliest stories of the living dead seem to have emerged from West Africa in the sixteenth century. These stories travelled with African-American settlers to America and the Caribbean islands. They were incorporated into voodoo practices in Haiti, where the zombie archetype mirrored the misery and brutal subjugation during slavery that existed there from 1625 to around 1800. Voodoo practitioners made concoctions from puffer fish toxins, mandrake juice and mushroom hallucinogens which put victims into a hypnotic state and sometimes into prolonged animated suspension. Zombification may be a state of catalepsy or motor paralysis induced by neurotoxins and is often followed by a long sleep, coma or even death.

There are several societies and clubs dedicated to zombies but perhaps the most authoritative and serious among them is the Zombie Research Society that has eminent scientists, film-makers, writers and public health experts as its members. They define the modern zombie 'as a relentlessly aggressive, reanimated human corpse driven by a biological infection', and they have imagined possible scenarios for the rise of zombies. 'This is our official position,' they say, 'and we maintain that nearly all aspects of the undead can be explained biologically and scientifically.' Reading through the Society's webpages suggests that the state of 'undeath' is triggered by environmental hazards, chiefly radiation, toxic chemical spills, viruses and spontaneous genetic mutations. According to the Society, zombies reproduce through consumption, not procreation, perpetuating a state of being removed from the provinces of birth and death, depletion and regeneration, illness and well-being. One of the defining traits of the living dead is their violent, insatiable hunger for human flesh. This behaviour is assumed to result primarily from their instinctual drive to infect others with the virus. By this definition, viruses themselves can be called zombies. Viruses, after all, come to life only when they infect a host. They alter the behaviour of the host and can even lead to the host's death.

Myths surrounding bloodsucking humans and creatures like bats have endured for over 400 years and have waxed and waned over time. They have appeared in waves in literature and in the movies too. The first vampire frenzy miffed the otherwise equanimous Voltaire so much that he wrote that 'nothing was spoken of but vampires, from 1730 to 1735'. Zombie is an established subgenre of horror and sci-fi in literature, graphic novels, animation and movies. According to Wikipedia, there have been at least 540 zombie films made in popular world cinema until August 2020. Hollywood has made the greatest share of zombie films, followed by Britain and Eastern European countries, and then there are a few from India and New Zealand, and single films made in Cuba, Egypt, Turkey, Indonesia, Singapore, the First Nations of Canadian Arctic, Pakistan and Vietnam. And not all zombie movies are out-and-out horror films. The 2006 Kiwi film, *Black Sheep,* had interesting comic bits, and the tag line at the end of the movie's trailer reads—*the violence of the lambs*. I presume you can imagine the plot.

In September 1998, the journal *Neurology*, a specialist publication on the subject, published a review by a Spanish physician, Juan Gómez-Alonso, on the impact of rabies on the functioning of the brain and the nervous system. He traced the origin of the vampire myth to some time between 1721 and 1728 in Hungary. Several newspapers and magazines wrote about Gómez-Alonso's paper, including the men's magazine *Playboy*, which seemed elated to report that some rabid male patients become like rabbits on heat with an overpowering desire for sexual gratification, just like vampires reawakened from their graves. Gómez-Alonso's paper finds interesting similarities between an archetypal vampire and a rabies-infected individual. Both conditions are transmitted through bites. In advanced stages of rabies, victims develop facial spasms, clench their teeth and retract their lips. For the faint-hearted who may not have seen horror movies, these resemble the characters of Michael Jackson's music video, *Thriller*. In much of Victorian literature, the lifespan of a vampire was said to be forty days, which is the same as the average time that it takes from the first symptoms of human rabies until death. Gómez-Alonso concludes that vampires and werewolves in historical accounts were rabid individuals, and their transmogrifying and facial distortions were construed as demonic by naive and god-fearing people.

This woodcut by the German artist Lucas Cranach the Elder titled 'The Werewolf or the Cannibal' (1510–1515 CE) depicts a rabies-infected man walking on all fours with a baby in his mouth, while body parts of other victims are strewn on the ground. A desperate and horrified woman and her children in a cottage are in the background, hiding from the rabid man.

Whatever the explanations about zombies and vampires, and even as the Zombie Research Society and enthusiasts prepare for Z-day, there are some microorganisms that infect and zombify creatures every moment of the day. In the scrublands and forests of Africa and South Asia lives a resplendent metallic green and wispy wasp no bigger than an adult's thumbnail. This is the jewelled cockroach wasp, a master hunter that uses the venom in its sting to stun insects and spiders. The female jewelled cockroach wasp has an iridescent abdomen and underbelly that stands out on the forest floor, though the purpose of her bright colouring is not quite understood. But any insect or spider that spots her would do well to steer clear of her. When she is pregnant, the jewelled cockroach wasp starts looking for a very specific host—a cockroach. She hunts it not for herself but for her eggs which she needs to deposit *inside* the cockroach. Catching a cockroach, however, is anything but simple on the forest floor, where they constantly scurry about from under one fallen leaf to another. When the wasp finds a forest cockroach, she lands right in front of him and injects him with a protein-shake of venom. The venom is fast-acting and stuns the cockroach almost instantly, paralysing its front legs and preventing it from fighting her. The wasp then injects a second venom directly into the cockroach's brain, and amazingly, this dose of venom brings the cockroach completely under the wasp's control. He (or she) now becomes docile, completely oblivious to any danger. The wasp leads her victim to an underground burrow and just before entering the burrow, she lays her eggs directly between the segments of the cockroach's body. She then conceals the burrow's entrance with soil and leaf litter. This is to prevent other predators from finding the cockroach or her eggs. After a day, the larvae of the wasp emerge and they spend the next few days sucking out the cockroach's body fluids before burrowing deeper inside and beginning to feed on the body parts of the docile forest roach. They eat the least essential parts first and save the nervous system and breathing system for last. This way, the living meal lasts about ten to twelve days, and during all this time the cockroach is alive and powerless to respond. The larvae continue to grow and develop until all that remains is a husk of the cockroach and from this empty exoskeleton emerge fully mature adult wasps, each female among them armed to repeat the gruesome cycle for itself. There are other parasitic wasps that deposit their eggs on other live hosts like a caterpillar of a moth, butterfly or beetle.

But how does a wasp's egg overcome the immune response of a cockroach or caterpillar? This remained a physiological enigma until the 1970s when scientists discovered that each wasp species harbours its own genetically unique virus from a family of viruses called *Polydnaviridae* (or the polydnaviruses, PDV), and it is the virus that enables the egg to get past the cockroach's defences. The PDV exists in two forms through the wasp's life cycle. Genes of PDV are integrated into the wasp genome where it remains dormant. But when the female wasp is ready to deposit her eggs, virions emerge from the wasp DNA and surround the egg, and these virions are injected into the caterpillar along with the eggs. This co-evolution of PDV with wasps occurred over the past 100 million years with the 18,000 or so parasitoid wasps that live today, and each type of wasp has evolved its own specific virus. Somewhere along this journey of mutualism and gene-swapping, the identities of the virus and wasp became blurred. And the integration is so complete that it is difficult to differentiate between the virus and its host. The genes containing the virus are passed on to the next generations.

There is a crucial difference between 'parasitic' and 'parasitoid' in nature—the former infects, breeds and then leaves, often without killing the host; the latter necessarily kills its host upon leaving. Some of those who first observed this behaviour of egg-laying wasps found it ghastly. A mortified Charles Darwin in an 1860 letter to his friend, the botanist Asa Gray, said, 'I cannot persuade myself that a beneficent & omnipotent God would have designedly created the *Ichneumonidæ* [the parasitoid wasp] with the express intention of their feeding within the living bodies of caterpillars, or that a cat should play with mice.' Others have found it inspirational. Legendary Swiss artist H.R. Giger, movie director Ridley Scott and writers Dan O'Bannon and Ron Shusett, created one of the most enduring extraterrestrial monsters on film based on the life cycle of the parasitoid wasp. In the classic sci-fi horror film *Alien* (1979), the alien is depicted as a terrifying insect-like creature protected by a silicon-based external skeleton, with acid blood running through its body that can

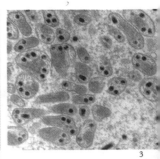

1   2   3

An emerald wasp attacks and paralyses a cockroach (1); a young wasp breaking out of a cockroach husk (2); and the polydnavirus (3), seen as black spots within the grey ovary cells of a wasp, which helps the wasp egg and larva evade detection and protects against getting neutralised by the cockroach's immune cells.

scythe through human flesh. The Alien mother is in search of a human host, and once she finds a way into a spaceship, she stings and paralyses the ship's crew, infecting them with her egg. When the egg hatches, it eats the crew from the inside, before bursting out in most gruesome ways—all of which is frighteningly reminiscent of the wasp's life cycle!

But even this horrifying phenomenon is not without its benefits (besides to the wasps, of course!). Agricultural scientists are studying the behaviours of different wasp species and their viruses that infect specific hosts in order to explore the possibilities of using this relationship to ward off insect pests. The tiny 2-millimetre-long, gnat-sized *Trissolcus* wasps found in temperate forests of East Asia, for instance, are proving to be effective in decimating the peach and apple orchard-loving brown marmorated stink bug in the US, thus saving billions of dollars' worth of produce.

Viruses are also deployed by trees in nature to protect themselves. Insect larvae are fleshy and full of protein and are found in abundance in spring. During this time, an explosion of very hungry caterpillars can strip trees of their leaves. But the caterpillars of butterflies and moths, the larvae of insects in general, are also among the most preferred hosts of parasites, and some forest trees and shrubs have developed an evolutionary strategy to protect themselves by attracting the enemy's enemy. Trees emit volatile organic compounds when caterpillars aggressively begin feeding on their leaves. Three chemicals in particular —benzyl nitrile, a mild cyanide poison; acetic acid, commonly known as vinegar; and 2-phenylethanol, which has a pleasant floral aroma and is commonly found in essential oils—are released in tiny quantities by tree canopies and these chemicals attract insects in large numbers. This may seem counterintuitive, but the gamble of attracting a large number of insects usually pays off for the trees because among the bevy of insects that arrive are a few that are already infected by a specific virus—the *Baculovirus*. In fact, these virus-infected insects are usually among the first to arrive, attracted by the aroma emitted by the trees. These insects shed baculoviruses on to the tree's foliage and all the tree has to do then is wait for a hungry caterpillar to accidentally ingest a baculovirus-laced leaf. When the baculovirus (Latin; *baculo*: rod or stick, like in bacteria) enters the caterpillar's gut, it multiplies rapidly before it spreads via the caterpillar's tracheal and circulatory systems into all parts of its body. The caterpillar becomes a factory for the virus, and eventually swells with billions of viral particles, ready to burst. Meanwhile, the insects that had initially arrived with their viral cargo also begin to die and the virus inside them then also

1 2 3

Once infected with baculovirus (1) an infected caterpillar (2) turns turgid and attempts to reach the highest parts of a branch. It begins to ooze fluids that release most of the virus particles from its abdomen. When it dies, the caterpillar's body is glued to a twig (3) where a passing caterpillar, beetles or spiders feed on the virus-laden body of the caterpillar, ingesting the virus.

finds release, spreading it farther. The virus manipulates the caterpillar's physiology to stop it from moulting and hijacks its brain to seek light, a phenomenon not seen in uninfected caterpillars. The heavy and turgid virus-laden caterpillars head laboriously for the treetop in search of light. When these obese, zombielike caterpillars reach close to the top, their tightened skins rupture in the middle with a pop and their dissolved innards burst open, releasing a spray of billions of virus progeny. These descend like tiny drops of gooey brown liquid, covering leaves and petals with small specks, for other hungry caterpillars to get infected with.

Not all the viruses are released when the caterpillar bursts open. Caterpillar cadavers containing viruses stick to leaves or waft away and get stuck to other leaves or fall to the forest floor. Here they are eaten by other creatures. The chemical make-up of the gut of insects and spiders is a little more alkaline than that of the caterpillar and this enables the viruses to survive inside them. Most predators too remain unaffected by the virus and release the viruses through their faeces which continue to reinfect new larvae of butterflies. The virus is sometimes carried away by the wind and in the gut of migratory insects and gets a free ride across relatively large distances, waiting to be gorged on by another hapless caterpillar.

Some viruses also help trees and shrubs to flower gregariously at a more favourable time and attract pollinators, thus improving the chances of survival of their hosts and themselves. A few recent studies have also shown that some viruses aid in plant survival especially under conditions of drought or severe cold. Scientists have found a species of grass growing in hot springs in Yellowstone National Park, and inside it lives a fungus, inside which lives a virus, and all three depend on each other for survival in these extreme, hostile conditions.

One of the most common and elegant examples of zombification is seen in ants. An ant colony is made up of millions of individuals and can dominate a hectare of forest or grassland. No creature other than those that feed exclusively on them can take chances with a fierce marauding army of ants. But thanks to a parasitic fungus called *Cordyceps* (Latin; *cord*: club; *ceps*: head), ants don't always have it their own way. Because of their fecundity, a potential parasite needs to infect only a few individuals who can, in turn, infect an entire ant colony. Spores from *Cordyceps* infiltrate an individual ant's body just before the onset of the monsoon. Thread-like fungal filaments ('mycelia') begin to grow inside the ant's body, first taking control of its legs and then, its brain. When humidity is high, its fungus-addled brain directs the ant to climb up the stem of a plant, just above the ant colony. The ant is utterly disorientated and entirely under the control of the fungus, and in its final act, it tightly grips the stem with its mandibles and legs, trembles and dies. From the head of the dead ant, a thin fungal strand emerges and grows for the next two or three weeks. At the end of each mycelium lies a bulbous 'club head' laden with tens of thousands of spores which bursts when the humidity is just right. If one of the drifting spores finds its way between the body segments of an ant, that ant, too, then meets the same fate as the dead ant. If this process goes on repeatedly unchecked for a month or two, the virulent fungus can wipe out an entire ant colony. There are at least 1800 species of *Cordyceps*, each specialising on parasitising just one or a couple of insects and arthropod species (spiders, scorpions, mites, centipedes, and so on). Despite the widespread annihilation, such epidemics have a positive effect on the jungle's diversity. Pathogens like baculovirus and *Cordyceps* prevent the dominance of any one creature (such as massive armies of ants) in any ecosystem for a prolonged period of time. The more numerous a species becomes, the more likely it is that it will be attacked by its microscopic nemesis and that more new pathogens will evolve to infect it.

We return now to the question of cannibalism. It is an integral part of zombie games and finds a place in proclamations of the Zombie Society. Cannibalism may involve eating one's own kind or others like themselves. It was once thought to be a rare occurrence in nature but is now recognised to be widespread among predatory and omnivorous animals, including traditional and sometimes even among modern human societies. But does cannibalism benefit a virus? Cannibalism by itself is

1

2

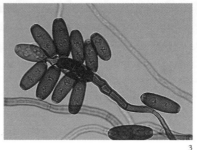

3

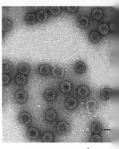

4

Geysers and hot springs in Yellowstone National Park (1) have roiling water gushing out through vents. Notice anything odd? While the trees and shrubs are keeping far away, a lawn of grass manages to grow along the edge of the geyser. The soil temperature that ranges between 35 to 65 °C is too uncomfortable for most plants. But not for this 'rosette grass' (2) or Dichanthelium lanuginosum var. thermale or, more commonly, 'panic'. Its name derives from the Latin 'panicum', referring to foxtail millet, and not from any botanical phobia. Panic is found commonly across woodlands of the Americas but what allows this grass to survive in such adverse conditions is a fungus (Curvularia protuberata) (3) that grows within its pith, and inside the fungal filament resides a virus (Curvularia thermal tolerance virus, CThTV) (4) that enables it to become admirably heat-tolerant. The temperature of the soil rises when a geyser erupts, and with each eruption CThTV pre-emptively prepares the fungus to produce more of the specific enzymes which protect the plants from extreme heat and water loss. On their own, neither the fungus nor the virus are able to tolerate temperatures above 38 °C, but together they are able to withstand much higher temperatures. When we think 'virus', we think only of their pathogenic associations. But in nature, their relationships are a lot more subtle and complex than we imagine. Scientists are exploring how beneficial viruses like CThTV and their fungal hosts can help crop and fruit plants to adapt to rising temperature caused by global warming. Next time you are in Yellowstone, look out for this ecological ménage à trois.

a poor mechanism for the spread of a disease. When a cannibal eats an uninfected individual, it kills a potential host for a disease, which is, of course, bad for the virus. When a cannibal eats an infected individual, there is always the possibility that it may remain uninfected. For the virus, one-on-one cannibalism may be a dead end unless more than one individual feasts on an infected victim (as in kuru). The chance of a virus to infect a huddle of uninfected or previously unexposed cannibals is therefore infecting all or none. Even this is not a good proposition for the virus because a virus needs a steady population of hosts that it can constantly reinfect after it is done replicating within one host. In some species of butterfly caterpillars, an infected caterpillar is smaller and becomes easy prey for a healthy caterpillar to cannibalise (Eric Carle's *The Very Hungry Caterpillar* may well be a horror film in the making). By eating a sick individual, only a few caterpillars become infected and most survive, and this reduces the spread of the baculovirus infection. Cannibalism, therefore, and viral infections, do not necessarily complement each other and the former is not a natural progression of the latter.

With or without cannibalism, the zombie myth is alive and kicking. After films, it is the gaming industry that has taken the zombie craze to the next level. A blockbuster PlayStation 4 game, *The Last of Us Part II*, which was launched, ironically, in June 2020 when America's COVID-19 numbers were at their highest, has as its protagonist a virus-mutated *cordyceps* that infects and zombifies humans. In the game, the human *cordyceps* outbreak occurs just outside the city of Austin, Texas in late September 2013, and within a few months it spreads to the rest of the United States, infecting or killing roughly 60 per cent of its citizens. The condition was termed 'Cordyceps Brain Infection' in which the fungus grows with the host still alive, with 'the Infected' (as the game developers label their zombies) undergoing four stages of infection and in some cases even zombifying. For those who do not want to get addicted to the game but would still like to construct their own zombie scenario, check out an excellent online simulation called 'Zombietown USA' that allows you, the user, to control variables such as the points of outbreaks and transmission rates (try it using early COVID-19 data!).

Are zombie scenarios at all realistically possible (I am serious)? Virologists, geneticists, evolutionary scientists, theoretical biologists and others will probably find it difficult to provide a definite answer. But while the idea of zombification is perhaps going a bit too far, in some ways investigating the virus's impact on the brain may be important because viruses do possess the incredible ability to shuffle genes and drive mutations in a very short span of time to create something entirely new. Even small tinkering by viruses can have widespread implications. A study published in *The Lancet Psychiatry*, for instance, has found that almost one in five people who have had COVID-19 develop some form of mental illness, like anxiety disorders, insomnia and dementia among others, within three months of testing positive. Even before the study, we had been warned by public health advocates of an approaching storm of mental health issues arising from the pandemic. Extreme biophobia and compulsive use of disinfectants may be just one small manifestation. The challenge for present and future human society will be whether or not we are able to grasp changes caused by viruses quickly enough to be able to respond effectively to them. Our experience with the COVID-19 pandemic suggests otherwise.

13

ENEMY'S
ENEMY

Say the word 'virus' and the first thought that comes to mind is of the
diseases they cause. But what if I told you that viruses can be, and have
been, a *cure* for some of the most intractable diseases known to man? Our
story begins at the River Ganga.

Between the seventeenth and nineteenth centuries, most European
nations sought to find riches in the Orient, the 'Dark Continent' and other
uncharted lands. In places far removed from their own cool temperate
climes, Europeans encountered new diseases, fevers and malaise, and they
sent forth legions of doctors and bacteriologists to safeguard their trade
interests from epidemics caused by hitherto unknown diseases. These
scientists were trained in newly established laboratories in European cities
and university towns, and they set up field stations in places as far apart as
Nha Trang in Vietnam in the east, and Havana, Cuba, in the west.

India, of course, was the jewel in Britain's proverbial colonial crown and
among the scientists sent out to India was Ernest Hankin, who arrived
at the port of Bombay in 1891. Hankin was a bacteriologist by training
who had matriculated from St John's College, Cambridge, in 1886, and
had interned in the labs of eminent scientists like Frank Wesbrook in

London, Robert Koch in Berlin and Louis Pasteur in Paris, before taking up his assignment in India. Hankin was appointed chemical examiner, government analyst and bacteriologist for the United Provinces, Punjab and the Central Provinces, and posted in Agra. His main task was to protect British troops from infectious diseases, especially the dreaded cholera. Cholera is caused by a sausage-shaped bacterium transmitted to humans through contaminated water or food and triggers violent diarrhoea characterised by 'rice-water stools'. If left untreated, cholera can be lethal. Much of Hankin's work entailed collecting water samples all year round so that he could monitor outbreaks of cholera in towns and villages. He was a keen swimmer and a skilled boatman and enjoyed taking boat rides along the Jamuna and Ganga when he went out on these inspections.

In February 1894, Hankin was on one such quest. He was on the lookout for the first signs of a cholera outbreak at the Maagh Mela in Allahabad, a fortnight-long Hindu festival that attracted over 3 million pilgrims every day. During the festival, devotees took a holy dip at the confluence of the mighty Ganga and Jamuna rivers, and remained on the riverbanks for days. Hankin collected and analysed water samples and what he found confirmed a pattern that he had noticed since he first began inspections of these rivers—there was very little bacterial contamination in their waters despite the multitudes of people and their cattle bathing in them, discarding their waste and burning corpses along their banks. In an 1895 paper, Hankin wrote that the Ganga and Jamuna were cleaner than most British or European rivers despite the way they were treated. Hankin knew about the Great Stink of London (1858) and Paris (1880) from his student days and wondered how and why these Indian rivers were able to avoid that kind of decay.

Local hakims and priests, of course, ascribed this to the mythical powers of the holy rivers but Hankin thought there might be a more scientific explanation. Was it possible that these waters possessed some kind of antibacterial properties? In the March 1895 issue of the *Indian Medical Gazette*, Hankin wrote about his 'Observations on Cholera in India', saying that he had sampled and tabulated each pond, well and water carrier along the two rivers—those that had the cholera bacterium and those that did not—and had found a 'protective substance' in the waters that did not contain the cholera bacterium. This substance could pass through a filter and its bacteria-killing property was destroyed when it was heated. He concluded that the filth and dirt thrown into the river were mostly consumed by turtles and vultures ('efficient undertakers', in his words), but that the remainder of the organic decay caused by rotting bacteria was quickly cleared by this unidentified 'anti-bacterial' material. He wrote about the antibacterial substance in the rivers in the *Annals of the Pasteur Institute* in 1896.

News of Hankin's observations reached Mark Twain, the famous American author of *The Adventures of Tom Sawyer,* who was on a tour of Europe and India with his family. When Twain reached Agra between 27 and 29 February 1896, he made it a point to seek out Hankin. 'A word further concerning the nasty but all-purifying Ganges water,' Mark Twain wrote after their meeting.

> When we went to Agra, by and by, we happened there just in time to be in at the birth of a marvel—a memorable scientific discovery—the discovery that in certain ways the foul and derided Ganges water is the most puissant purifier in the world! This curious fact, as I have said, had just been added to the treasury of modern science. It had long been noted as a strange thing that while Benares is often afflicted with the cholera she does not spread it beyond her borders. This could not be accounted for. Mr. Hankin, the scientist in the employ of the Government at Agra concluded to examine the water. He went to Benares and made his tests. He got water at the mouths of the sewers where they empty into the river at the bathing ghats; a cubic centimetre of it contained millions of cholera germs; at the end of six hours they were *all dead*. He caught a floating corpse, towed it to the shore, and from beside it he dipped up water that was swarming with cholera germs; at the end of six hours they were *all dead*. He added swarm after swarm of cholera germs to this water; within the six hours *they always died*, to the last sample. Repeatedly he took pure well water which was barren of animal life, and put into it a few cholera germs; they always began to propagate at once, and always within six hours they swarmed—*and were numberable by millions upon millions*.

> For ages and ages the Hindoos have had absolute faith that the water of the Ganges was utterly pure, could not be defiled by any contact whatsoever, and infallibly made pure and clean whatsoever thing touched it. They still believe it, and that is why they bathe in it and drink it, caring nothing for its seeming filthiness and the floating corpses. The Hindoos have been laughed at, these many generations, but the laughter will need to modify itself a little from now on. How did they find out the water's secret in those ancient ages? Had they germ-scientists then? We do not know.

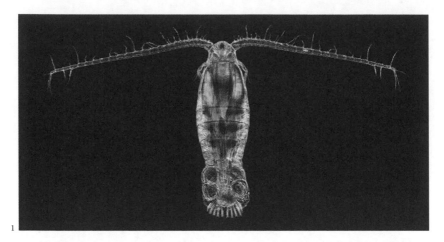

1

Cholera is a bacterial disease that occurs when an ingress of seawater or bacterium-tainted water mixes with freshwater. The bacterium Vibrio cholerae *is found naturally in seawater, and it spreads more efficiently when seas are warm and support their multiplication inside the bodies of small marine arthropods called copepods (1). Although widely prevalent across south and south east Asia, cholera is believed to have originated in the intertidal regions of the Ganges delta, and became pandemic between 1817 and 1824. It was spread across continents through the ballast water of ships and in the bowels of sailors who released the bacterium in port cities. Once the bacterium reached new shores, it gradually evolved into new local strains. Interestingly, although the bacteria* Vibrio cholerae *(2) is mildly toxic to humans, it becomes extremely pathogenic, infecting humans and producing characteristic cholera 'rice water stools' due to the cholera toxin bacteriophage phi or CTXφ (3). As with any phage, CTX needs to take control of the bacteria to propagate itself. In exchange for this control, however, the phage confers to the* Vibrio *the cholera toxin that makes it more efficient in infecting human hosts. CTX phages have repeatedly helped* Vibrio *to acquire new genes and to develop into new strains. This means that the* Vibrio *is ever-evolving, and therefore, there is very little hope*

Hankin's findings on the antibacterial properties of the Ganga's waters were published in reputed journals, but despite such unimpeachable endorsements, his work remained largely unnoticed. It didn't help that Hankin also suffered from self-doubt. He saw several limitations in his methods and could not quite conclusively explain what caused the waning of the cholera bacteria in some wells but not in others. Hankin, therefore, may have been the first to observe the phenomena of viruses that kill bacteria but his name remains little known outside a small circle of virologists, and during his tenure in India, his best advice to stave off cholera in British cantonments and colonies continued to be to use antiseptic potassium permanganate. He advocated adding this purple chemical to wells and instructed army cantonment officers and new officers to India to eat 'purple salad' to avoid diarrhoea.

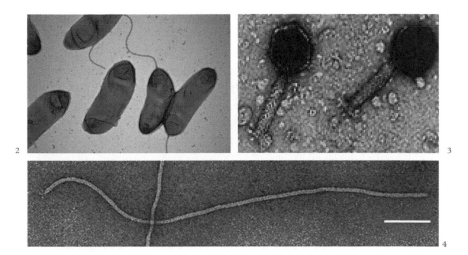

of developing an effective cholera vaccine. The cholera Vibrio *is not alone in benefiting from phages to turn pathogenic; other bacteria like the ones that cause scarlet fever, diphtheria, botulism, diverse forms of food poisonings, and lung infections, all produce potent exotoxins (exo- to release outwards) with the help of phages. Some of these phages do not have the versatility of CTXφ and therefore it is simpler to make effective vaccines against them (e.g. diphtheria). In an additional twist to the tale, while one bacteriophage bestows added virulence to the* Vibrio *bacterium, there are a few vibriophages like Vp5virus (4) which attack and kill the cholera bacterium and these phages can also cure mild to severe cases of infection. These two different types of viruses, one which spurs the disease, and the other with the ability to kill the pathogen, are often found residing in the same water. For cholera to become an epidemic, the bacterium needs several things going for it. First, it needs warm waters where the right copepod species or algal blooms are nourished that will host the* Vibrio *and the phage. Second, it needs a warm high tide to take the seawater inland, where it will contaminate a drinking water source. And third and most importantly, it needs to infect a malnourished and susceptible population who will carry the vibrio in their gut, and taint other ponds and wells. Sea level rise and frequent floods caused by global warming will be the ideal conditions to proliferate copepods and* Vibrio, *thus potentially increasing the number of cholera outbreaks.*

Over the next fifteen years, although the phenomenon of bacterial destruction by unidentified agents was observed by other scientists in Europe, no one pursued it doggedly to its logical end. Then, in 1912, Frederick W. Twort, a biologist who was attempting to grow the smallpox virus using calf's lymph, noticed that something induced a watery dissolution of the bacterial colonies where he was culturing the lymph extract. He observed a 'glassy transformation' in the Petri dishes that had bacteria growing on them. When he separated these circular glassy discs and attempted to grow the bacteria again, he failed. He concluded that these bacterial cells had been destroyed by a 'filterable agent' which he believed were viruses.

Around the same time, at the Pasteur Institute in Paris, completely unaware of Twort's and Hankin's work, and under very different circumstances, a Canadian émigré, Felix d'Hérelle, discovered a 'microbe' that was 'antagonistic' to bacteria—that is, it caused the bacteria to die, creating discrete circular patches or 'plaques' on the surface of the Petri dish on which he was growing a lawn of bacteria. D'Hérelle concluded that the invisible microbes that were causing the bacteria to die were 'ultraviruses' that attacked the bacteria and multiplied at their expense. And it was he who labelled these viruses 'bacteriophages'—eaters of bacteria.

D'Hérelle was a largely self-taught bacteriologist who, because of his excellent skills with creating growth media for bacteria, had been asked by the Pasteur Institute to be part of a team to control an outbreak of dysentery among French soldiers and was later sent on an assignment to stop diarrhoea in hogs. It was during this time that d'Hérelle spotted the plaque (he called them *taches vierges*, 'virgin spots') that formed when dysentery-causing bacteria were killed by the 'bacteriophage'. D'Hérelle began to use bacteriophages to cure dysentery and other stomach infections. He observed that after killing the disease-causing bacteria, the virus or the bacteriophage did not persist in the system. It, too, disappeared along with the bacteria. D'Hérelle summarised the action of bacteriophages eloquently: 'In convalescent cases . . . the antagonistic microbe vanishes soon after the pathogenic bacillus disappears . . . I have never seen or isolated this antagonistic microbe in [my] subjects.'

All studies confirmed that the antagonistic action of the invisible 'antibacterial agent' or bacteriophage was seen only in *association* with that specific bacterium. In other words, the agent was host-specific. When d'Hérelle swapped his agents, using, for instance, a paratyphoid fever-destroying agent on dysentery-causing bacteria, he found that there was no action. D'Hérelle published his preliminary findings in 1917 but did not cite the works of either Twort or Hankin in his paper. D'Hérelle's thesis on phages contradicted the position taken by some eminent bacteriologists of his time, including Jules Bordet, who was awarded the Nobel Prize in 1919 for his work on immune responses and serum components. Bordet and his co-workers believed that the destruction of the bacteria was caused not by 'a particulate ultramicrobe' but a bacterial *enzyme*, and that the invisible agent that d'Hérelle wrote about was actually self-restricting enzymes released by neighbouring bacterial colonies—a bit like 'allelopathy', a phenomenon where some plants are able to secrete chemical substances that prevent or inhibit other plants from growing around them.

Bordet was an imposing figure in his field. He had founded the Pasteur Institute in Brussels in 1901 and remained its director until 1940, while d'Hérelle was a mere volunteer at the Pasteur Institute in Paris with no standing in the scientific world. Bordet's views naturally carried more weight and strongly influenced the opinions of the scientific community, undermining the credibility of d'Hérelle's interpretation. Bordet also used Twort's 1915 paper in *The Lancet* on bacterial transformation to discredit d'Hérelle as the discoverer of the phenomenon. Twort, meanwhile, caught up in his wartime responsibilities and post-war research, was initially unaware of the fracas in France. Had Bordet not invoked Twort's 1915 paper as a challenge to d'Hérelle, Twort, too, might have been relegated to an insignificant footnote in the history of science. But the Twort-d'Hérelle Controversy (as Parisian papers called it) carried on for more than ten years and Bordet brought Twort on board to support his arguments.

D'Hérelle, meanwhile, continued to demonstrate the efficacy of this yet unseen but powerful agent in the fight against infectious disease. After World War I ended, d'Hérelle proved the efficacy of 'phage therapy' by curing a twelve-year-old boy suffering from bacterial dysentery at the Hôpital des Enfants-Malades in Paris. He successfully treated his young patient with a phage isolated from a faecal sample taken from another dysentery patient. He ingested the phage preparation himself to test its safety before administering it to the boy (it was common practice at the time for researchers to self-test, before modern clinical trial procedures were widely accepted). Over the next three weeks, he cured three additional dysentery patients using the same phage preparation. Other scientists across Europe began to take note of his results. In 1926, d'Hérelle treated four cases of bubonic plague with an anti-plague phage in Alexandria on his way to Bombay. The French medical periodical, *La Presse Médical,* reported on this work and cemented d'Hérelle's reputation by the time he arrived in India. He travelled widely in India and initiated field trials for cholera in Calcutta, Patna, the Punjab, Bengal and Assam, and for the plague in Hyderabad and Agra in 1926. Phages isolated from (local) cholera victims, it was claimed, ended the cholera epidemic within forty-eight hours, as compared to conventional interventions which achieved the same result only after twenty-six days. However, this was the period of Mahatma Gandhi's satyagraha (*satya*: truth, *agraha*: appeal; a peaceful civil disobedience movement) which led to 'non-cooperation' by Indians with the British government, and despite encouraging early results, the trials were disbanded.

In 1932, a three-scientist commission in Paris decided to compare Twort and d'Hérelle's material and concluded that they presented the same phenomenon but credited d'Hérelle with the discovery of phages because he had gone two steps further. D'Hérelle had not only isolated the agent, he had also shown that it caused repeated killing of bacteria even after dilution. He also exhibited that this phenomenon was observed in other human and animal diseases as well. It had taken a very long time, but eventually, the scientific community reluctantly, and somewhat sheepishly, recognised d'Hérelle as the discoverer of bacteriophages.

While discussions and experiments on the efficacy of phages were taking place in Europe, and scientists and physicians in a few clinics in Paris, Warsaw and St Petersburg were dispensing treatments for specific diseases, there was very little uptake of phage therapy across the Atlantic. Then, in November 1931, Tom Mix, a gunslingin' Hollywood star fell ill. It is likely that only diehard fans of Westerns will remember his name today, but many Beatles fans would have seen him (without recognizing him) on the iconic cover of Sgt Pepper's Lonely Hearts Club Band, where Tom Mix rubs shoulders with seventy-one other matinee idols, politicians, gurus, singers, artists, a boxer, Albert Einstein, and a few other oddballs. Mahatma Gandhi and Jesus Christ were removed at the last minute before the cover went to print. If you happen to have the cover of the LP, look for a man in a wide-brimmed Stetson Cowboy hat. He stands to the left of a technicolour John Lennon, between Oscar Wilde and Marlon Brando. That is Tom Mix—the greatest Cowboy star of the silent movie era of the 1920s and '30s. Mix had popular radio shows in English and Spanish, several dedicated comic books and cereal box labels. His gravity-defying stunts with Tony the Horse saved towns from bandits, warded off armed train robbers, and fought off marauding 'Indians' on wild broncos. Mix was brawny and handsome with thick black hair combed back and a deeply cleft chin. His popularity as a Cowboy star was such that he served as pall-bearer at the funeral of the legendary Western lawman, Wyatt Earp, in 1929.

Late on the night of 23 November 1931, after shooting at Universal Pictures Studio in Los Angeles, Mix collapsed, clutching his stomach. By the time he was brought to Hollywood's Clara Barton Memorial Hospital, his inflamed appendix had ruptured, spreading millions of bacteria in his stomach and leading to a critical condition called 'peritonitis'. News about Mix's illness spread fast. Nearly 4000 letters and 1500 telegrams poured into the hospital each day, and the likes of Harold Lloyd, Jack Dempsey, Babe Ruth, Will Rogers, Mexican superstar Leo Carillo and

Tom Mix, in a large white Stetson hat, stands to the left of John Lennon, dressed in a yellow costume.

director John Ford, sent in heartfelt wishes for his speedy recovery. Major newspapers on both coasts of the United States tracked his progress. Mix's doctors had very few medical options with which to fight his infection. The standard treatment for peritonitis was to insert a tube and rinse the abdominal cavity with a salt solution and then drain it—an intervention that was not likely to be much good in Mix's condition. Doctors at the Hollywood Hospital were so sure that Mix would not survive that they prepared his death certificate—*Thomas Edwin Mix, signed Drs Seroggy, Smith and Dennis, dated 23 November 1931 8 p.m., Diagnosis: Ruptured Gangrenous Appendix, Immediate Cause of Death: Peritonitis.*

But then someone in Mix's medical team thought of phages. Phage therapy was still in its infancy in the US, although a few pharmaceutical companies like Eli Lilly, E.R. Squibb and Parke-Davis had begun marketing phage preparations for treating urinary tract infections and dysentery. Mix's team reached out to E.W. Schultz, professor of bacteriology at Stanford University, who had only recently established a phage lab. For about $2, the lab in Stanford could prepare phages that were specific to some particular bacterial strains. On the afternoon of 23 November, Mix's doctor drained the excess infected fluid from Mix's stomach and sent a sample of the disease-causing bacteria to Schultz's lab so that they could concoct an exact phage match. Twenty-four hours later, at 6 p.m. on 24 November, a small propeller plane touched down outside Glendale Airport's Spanish-style terminal building carrying a supply of fresh phages from Stanford. Doctors injected the phage directly into Mix's stomach via a catheter and hoped for the best. Over the next four days, three more batches of cocktail phages arrived at the hospital, and on Saturday, 28 November 1931, *The New York Times* reported: 'Tom Mix Rallying Slowly'. Two days later, the doctors said the actor was 'brightening up', and on 3 December, a picture of Mix appeared on the front page of the *Los Angeles Times*, being fed his first solid food in eight days by a smiling, bonneted nurse.

Despite this rather flamboyant success, phage therapy did not quite manage to find its deserving place in the limelight. Most papers did not even mention the word 'phage', attributing Mix's recovery instead to a 'serum'. And even d'Hérelle, the world's leading authority on phages who was by now a professor at Yale and who seldom let slip an opportunity to tout the efficacy of phage therapy, did not leverage Mix's cure.

There had been other opportunities for phages to become better known in the US. In October 1930, Harry Sinclair Lewis became the first American author to get the Nobel Prize for literature 'for his vigorous and graphic art of description and his ability to create, with wit and humour, new types of characters'. His 500-page novel *Arrowsmith* was perhaps the first real science novel that revolved around the discovery and use of a bacterium-eating virus, and the complications of animal and human experimental research, morality and ethics in science. Lewis was introduced to the idea of phages and the miracles they had performed in Europe by his friend Paul de Kruif, a scientist at the Rockefeller Institute and author of a book called *The Microbe Hunters* in 1926. But Lewis's bacteriophage in *Arrowsmith* was probably mistaken by his readers as something fantastical out of science fiction and phage therapy remained at the fringes of medical intervention in the US.

In Europe, meanwhile, a few institutes and scientists continued to experiment with phage therapy. Among them was a friend and former colleague of d'Hérelle at the Pasteur Institute, Georgyi (or George) Eliava.

Eliava was an innovative bacteriologist in his own right, who returned to Tbilisi in the Soviet Republic of Georgia and founded the George Eliava Institute of Bacteriophages, Microbiology and Virology in 1923 which started as a small lab, and in 1937, with the help of d'Hérelle, became among the first state-of-the-art bacteriophage research institutes.

In the interwar period, especially between 1919 and 1930, the use of phages in medical treatment advanced to some degree, with short clinical reports appearing in medical journals. These successes were achieved despite no one still knowing exactly what bacteriophages were. At the same time, many in the scientific community continued to question the efficacy of phage therapy. Most phage therapy was done in desperate situations on desperate people and there were few meta-analyses or compilations of cases. Another reason why d'Hérelle's clinical trials were thought not to meet accepted standards was their lack of experimental rigour. D'Hérelle did not include appropriate placebo controls in his studies, which are regarded as de rigueur for all clinical trials. Perhaps he did not want to deprive any of his subjects of the benefit of his treatment and was convinced that his phage therapy could cure everyone. In prominent institutions like the Pasteur Institute in Paris or Brussels, phage research was accorded a low priority and very few scientists experimented with viruses. Some practitioners also recorded inconsistent results because they treated *all* symptoms with the *same* phage isolate, not realizing that phages are highly host-specific and that the therapy works *only* after identifying the pathogen. In the end, the small-scale phage trials were poorly documented and reported and did not hold up under close scientific scrutiny. Phage therapy was considered a miracle but was not fully accepted as science because neither could the agent be seen, nor the phenomenon be properly explained. It didn't help phage's case that hundreds of clinical reports appeared only in local journals, chiefly in France, Germany, Eastern Europe, Russia and even Brazil. In comparison, there were few studies in the US and the UK, and the English-speaking world, which rose to dominate the global order, never accorded prominence to the success stories of phage therapy.

Phages, however, continued to be used by desperate people. By World War II, although the first antibiotics had been discovered, very few countries had them, and it was phage therapy that saved countless lives in countries that did not have access to antibiotics. The Soviet, Polish and German armies used phages for treating wounds and diarrhoea. In Brazil, for over a decade (1929–40), the Oswaldo Cruz Institute deployed phage therapy to treat outbreaks of dysentery and developed effective phage combinations to fight staphylococcal infections. After World War II, antibiotic use became more widespread and antibiotics quickly became the mainstay of global public health. Post the War, the growing popularity of antibiotics among military doctors, and the euphoria of new arms and chemical technology, which were backed by the military-industry complex in the US, UK and Switzerland, gave these new medicines a big push. Over the next three decades, governments and corporations invested heavily in research and many new antibiotics were added to the

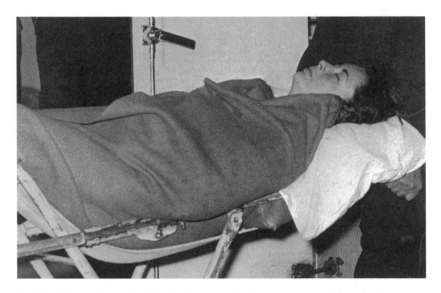

In 1961, Hollywood star Elizabeth Taylor was in London taking time off from the filming of the costume drama, Cleopatra. On the night of 3 March, Ms Taylor suddenly fell ill with suspected pneumonia and had to undergo an emergency tracheotomy. The press followed every little development as her condition deteriorated. On 7 March, a Philadelphia-area newspaper reported that a local company, Delmont Laboratories, had sent on the next jet to London 'an unusual antibacterial treatment' at the request of Ms Taylor's doctors. This treatment was staphylococcus bacteriophage lysate and was used in conjunction with a newly synthesised form of penicillin called 'methyl penicillin'. Ms. Taylor's condition gradually improved but it was not clear what had contributed to her recovery. On 12 March 1961, The New York Times *headlined, 'Widespread Germ Succumbs to a New Synthetic Penicillin' and made no mention of the use of the phage. Phage products of small companies such as Delmont were not approved as drugs until the late 1960s when they were also brought under strict regulation by the US Food and Drug Administration. With no incentives or investment for research as well as regulatory uncertainty, companies like Delmont had to stop selling phage treatments for human use by the early 1990s. The company continues to make a USDA-licensed phage therapy to treat dogs with recalcitrant staph infections on their skin. Ms. Taylor, who championed the cause of HIV/AIDS patients in the late 1980s and early 1990s, probably never even knew that it was a virus that may have saved her life. Two recent studies have shown that strain-specific phages overcome the deadly multiple drug-resistant* Staphylococcus aureus (MRSA) *infection more effectively than a combination of antibiotics.*

health arsenal. More than 120 new sources of antibiotics were discovered between 1930 and 1980, although only a dozen or so were adopted for public health programmes.

In all this time in the US and the rest of the world, there was virtually no funding for phage research. The only institute other than the George Eliava Institute of Bacteriophages, Microbiology and Virology, that sustained research on phage therapy was the Warsaw-based Hirszfeld Institute in Wroclaw, Poland, which was founded in 1952. America's leading medical association, the American Medical Association, presented inconclusive reports on the effectiveness of phages and many members favoured antibiotics over the use of phages. There was also general distrust, religious bigotry and limited understanding of the science behind using a virus to fight another microbe. Moreover, there was growing suspicion at this time in the West of anything Russian and Eastern European, and strong fear that germs might be used for biowarfare against the West.

Most antibiotics are bioactive chemicals that have been identified, isolated and extracted from soil bacteria or fungi, which kill a wide range (a 'broad spectrum') of human pathogens. In order to make these antibiotics more durable and stable, synthetic analogues are developed, after which the antibiotics are mass-produced. The way in which antibiotics 'work' inside a human wound or in conditions of fever is like that of a predator feeding on a susceptible pathogen—in principle, therefore, not very different from the pathogen-destroying action of phages themselves. But antibiotics were money-spinning endeavours and there was a race among pharmaceutical companies to produce more to capture domestic and global markets. For these companies, phage therapy was a less exciting proposition—unlike early antibiotics, which were presumed to be 'one size fits all', phages were complex to make and strictly strain-specific. Before good storage practices were developed, the short shelf life of phages and other storage and handling limitations were also seen as inconveniences. It didn't help that publications about phage therapy in influential journals dwindled, and although d'Hérelle was nominated eight times—every year from 1925—for the Nobel Prize, he never won it. Some historians of science have suggested that d'Hérelle winning the Nobel Prize could possibly have given phage therapy the scientific validation and public support it desperately needed.

By the 1960s, we looked unbeatable with our arsenal of antibiotics, and we got a little cocky. Western pharmaceutical companies and their foundations were convinced that with their growing arsenal of antibiotics, the end of the age of infectious diseases was in sight. The US Surgeon General, William Stewart, declared in 1967 that it was time to 'close the book on infectious disease'. In this age of ebullient optimism, a popular medical textbook, considered a classic in its day, went so far as to claim that all infectious diseases would be eradicated by the year 2000. Cadres of young medical students and practitioners were made to believe that there were now more pressing emerging concerns in health and medicine that they needed to focus upon. With new advancements and more individualised treatments, new hierarchies emerged in medicine. Life sciences like bacteriology began to play a supporting, rather than an integral, role in the cure of diseases. The study of disease outbreaks, new pathogenic agents and conditions in which they emerged, were largely ignored. Leaders in the field of the life sciences from the eighteenth to early twentieth centuries—men like Jenner, Koch, Pasteur, and others on whose discoveries the principles of modern medicine rested—still had their portraits hanging in the corridors of medical schools but their contributions became inconsequential miscellany, mere footnotes in the chapters of medical textbooks. From once being part of a vibrant, inspiring and networked community of scientists, medicine became insular, with narrow specializations. Backed by vaccines and a portfolio of antibiotics, we were lulled into believing that the eradication of several diseases was now only a matter of time. Year after year, investments and political commitments to fight infectious disease began to dry up and fewer new drugs aimed at diseases that affected the poorest of the poor made it to the market. As globalization and the exploitation of new resources occurred, new infectious agents did begin to appear, but these were treated as exotic fevers, titillating researchers and journals who wanted to claim these as their discoveries, and no serious plans were made to counter any potential threat. Unfortunately, the WHO and other UN agencies too, were caught up in this mindset.

In the space of a few decades, this rose-tinted picture began to acquire dark hints of foreboding. In his Nobel Prize acceptance speech in 1945, Sir Alexander Fleming, discoverer of penicillin, had presciently warned about the dangers of its misuse. Just twenty years later, reports of 'resistance' to penicillin began to trickle in from hospitals around the world. But we faltered at the start and kept fumbling in this race even as new antibiotics were being added, and what had seemed like a 100-metre dash against disease began to look like a gruelling, uphill marathon. By the early 1960s, we had begun using antibiotics widely and indiscriminately—in hospital settings and outside—often without prescription. We used it to speed up the growth of our livestock, and unscrupulous pharma companies sold antibiotics to unsuspecting poultry farmers promising their efficacy in averting potential disease outbreaks. In 1969, the WHO and the UN's Food and Agriculture Organization (FAO) laid down standards of permissible limits for antibiotics and other chemicals in food and water, thereby tacitly agreeing that such pernicious overuse was acceptable. Similar or related families of antibiotics that were being used to treat humans began to be used for livestock and poultry and this complicated matters further. One estimate suggests that since the 1990s, we have been using at least 1,20,000 tons of antibiotics annually worldwide, out of which well over 70 per cent is used as livestock supplements. As a result of decades of intensive chemical use in agriculture and other sectors, all foods we consume today contain traces of antibiotics and synthetic chemicals which cannot be degraded by microbes or by any other natural means. The intermixing of sewage and agricultural run-off has seeped into the food chain and allowed microbes in soil and groundwater to develop resistance. And it is not at all surprising that we began to lose several important battles to microbes.

Some of the first signs of antibiotic resistance appeared in the malarial parasite. It was then seen in gastrointestinal diseases like *Shigella*, and

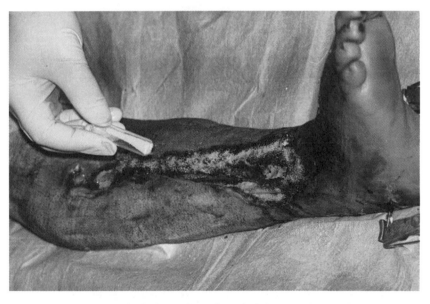

A wound infected with multiple drug-resistant bacteria being treated at the Georgia Technical University Hospital using a customised cocktail of bacteriophage (seen in the small vial in the hand).

gradually tuberculosis and other more widely prevalent diseases began to become drug-resistant. The overuse and consequent failure of antibiotics has developed into a major public health crisis now. Drug-resistant diseases threaten every country in the world and pharmaceutical companies are scrambling in their search for newer antibiotics to counter these. According to the US Centers for Disease Control and Prevention (CDC), more than 2 million people in the US become ill with antibiotic-resistant diseases every year and this has led directly to more than 23,000 deaths.

The overuse of antibiotics has also converted some otherwise benign bacteria into deadly pathogens. *Acinetobacter baumannii*, for example, is an opportunistic bacterium that invades the lungs and skin of immune-compromised individuals, especially those who are hospitalised for prolonged periods. Until the 1970s, *A. baumannii* was seen as a low-risk pathogen that caused little harm. But it has rapidly evolved to become a high-risk pathogen now and finds itself among six top-priority microorganisms that the Infectious Diseases Society of America (IDSA) is worried about. This bacterium is colloquially dubbed 'Iraqibacter'

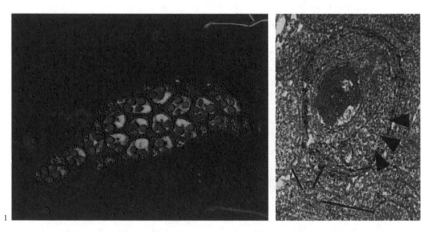

Bacteria belonging to genus Wolbachia *(1) (the round blue structures within the green ovary sac of an* Aedes aegypti *or the dengue mosquito) naturally occur in more than 65 per cent of all insect species and also in spiders, mites and some worms. What makes* Wolbachia *special is a minute spherical WO bacteriophage (2, marked with arrows) that lives inside it. Different species of this bacteria-bacteriophage duo play different roles in the variety of hosts they infect. For example, in a few insect species,* Wolbachia *helps to produce and process nutrients. In insects like butterflies,* Wolbachia *kills the males and in woodlice it feminises them. In most insects, however, the bacteria-virus duo affects its host in two ways. First, it makes the host populations less fecund. And then it prevents other bacterial pathogens from reproducing inside the hosts' bodies, thereby protecting them against other viral and bacterial infections. This multifarious male-killing, gender-bending, gonad-eating*

because soldiers who returned from Iraq were frequently found to be infected with it. The WHO predicts that deaths from resistant bacteria will rise from roughly 7,00,000 a year today to nearly 10 million by 2050, with significant contributions from previously non-pathogenic organisms such as *Acinetobacter*.

Another example comes from a common, yet deadly, bacterium called *Staphylococcus aureus*. Under a simple microscope, a Gram-stained slide of *S. aureus* appears like a pretty bunch of purple grapes. It is a common resident of the air and occasionally of water too, and we carry it on our skins and inside our lungs. It is practically ubiquitous and there is no reason why it should not be found floating inside hospitals. Until the late 1980s, *S. aureus* could easily be subdued by a wide range of antibiotics. Not any more. Today, if *S. aureus* lands on a patient who has suffered

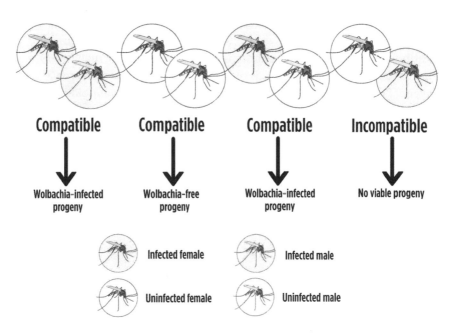

Compatible → Wolbachia-infected progeny

Compatible → Wolbachia-free progeny

Compatible → Wolbachia-infected progeny

Incompatible → No viable progeny

Infected female

Infected male

Uninfected female

Uninfected male

duo provides a ray of hope for eliminating the transmission and spread of vector-borne diseases. Mosquitoes which carry the virus-bacterium duo can potentially reduce the transmission of other pathogens to a population, as well as reduce or even eliminate local mosquito populations, thus stopping fevers like dengue and chikungunya. Scientists have successfully used this idea to reduce the numbers of dengue-spreading mosquitoes in towns of northern Australia, Nouméa in New Caledonia, and Yogjakarta, Indonesia; and Zika in some towns in Brazil. Interestingly, recent genome analysis of WO has found other animal genes in it, including a toxin gene very similar to what makes the bite of the black widow spider so deadly. How so many animal genes got into WO's genome and what purpose they serve is a genetic conundrum waiting to be deciphered.

burns or is recovering from post-operative surgery, it causes rapid infection, high fever and systemic complications, and if poorly managed, death ensues. *S aureus*' genes are now stubbornly resistant to all the antibiotics we know today.

The evolution of antibiotic resistance is a textbook case of what is called the Red Queen Effect (RQE) in evolution. This name derives from Lewis Carroll's *Through the Looking Glass,* where Alice is told by the Red Queen: 'It takes all the running you can do, to keep in the same place.' Like being on a treadmill—speed it up and you need to run faster and faster just to stay in the same place or risk falling flat on your face. When applied to evolution, the RQE means that animals or plants need to evolve faster to meet rapidly changing conditions or risk going extinct. Acquiring antibiotic resistance is the RQE of bacteria, and it was only a matter of time, no matter how careful we were, before antibiotic resistance among target and non-target microbes would emerge. Perhaps we could have delayed the moment if we had been more careful about our use of antibiotics, but antibiotic resistance is almost inevitable. Virtually every bacterium that infects humans is resistant to at least one antimicrobial agent. Those bacteria which develop resistance to a single antibiotic are simply called 'antimicrobial resistant' (AMR) strains. Far more dangerous are those that have developed resistance to multiple or *all* known antibiotics that would ordinarily kill them. They are labelled 'multiple and extremely drug-resistant' (MDR/XDR) strains.

We know that bacteria and viruses readily swap genes by horizontal transfer and there is free-floating genetic material that also accidentally gets incorporated into the genes of microbes. We therefore find antibiotic-resistant genes in the most unexpected places and in the most unexpected organisms.

There are no life forms without their specific viruses, and every species of bacteria possibly has a very specific phage that kills it. This is why phage therapy is such an important field to explore further and develop. There are effective phage alternatives that can help us, but they are underused and remain largely unknown to medical practitioners. So this is how phage therapy works—a cocktail of phages is used to target a wound in which drug-resistant bacteria are concentrated. These phages begin multiplying using the genes of the bacterium, the bacterium's cell proteins and energy. The more bacterial cells there are infecting the tissue, the greater is the multiplication by phages, thereby killing only the specific pathogenic bacteria until there are no more left. The microbiome per se is left unaffected. Virologists rather descriptively term this 'auto-dosing'—the phage produces as many of itself as the level of infection requires. Autodosing prevents the need for repeated administrations of phages to the treatment site. The danger of *over*dosing is also obviated because excess phages find no more bacteria to infect and are therefore cleared and ushered out by blood cells.

In some ways, phages are the epitome of *targeted* medicine and, therefore, have a clear edge over scattergun broad-spectrum antibiotics. Most phages infect only a few bacterial strains, leaving all other strains intact and unaffected. So naturally occurring microbes that otherwise do not harm or may even be of benefit to the host remain unaffected. Because the evolutionary relationship of phage and bacteria is historic, the possibility of the bacteria evolving and overcoming the phage's predation is highly unlikely. Phages can infect and kill all bacteria—whether they are natural, evolved or resistant strains.

Isolating a new phage is faster and cheaper than synthesizing a new antibiotic. In addition, rapid genetic sequencing techniques can prevent 'lysogenic' phages i.e., phages that incorporate themselves within the (bacterial) genome and avert the threat of passing their genes on to bacterial cells. Some treatments also harvest back the viruses that have emerged after killing the bacteria and curing the patient. So, if you were to use this therapy, you may well be able to provide (through harvesting) the phages to cure another patient. The phage must kill the bacteria through the lytic process—that is, it enters, takes over the bacterial DNA, makes several copies of itself and bursts out, killing the bacteria in the process. The virus therefore destroys its hosts and produces several clones of itself in turn. This behaviour of the virus reminds us of the demon in Hindu mythology, *Raktabīja* (*rakta* = blood, *bīja* = seed) who had infinite power to replicate. If a drop of his blood fell on the ground, it would spontaneously give birth to another of its kind.

By diving deep and searching the soil, the sea floor, our own bodies and even sewage passageways, for phages and their genomes, researchers are finding new species and varieties. The diversity of the world's bacteriophages is an essentially inexhaustible source that we can scour for antimicrobial treatment using modern metagenomic methods. The story of the neglect of phage therapy in modern medicine is similar to the way we have overlooked renewable and cleaner alternatives for energy. Phage therapy has only survived in a few institutions like the Eliava Institute in Georgia and the Hirszfeld Institute of Immunology and Experimental

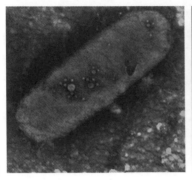
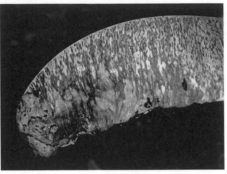

Phage therapy has tremendous potential to fight massive and devastating crop epidemics. The bacterium Ralstonia solanacearum *(1) lives on the surface of soils in heavily worked agricultural fields, and easily rides on dust. It also spreads by raindrop splatter and is carried by water. It causes widespread epidemics like these yellow and black streaks on banana leaves (called Moko [2]) leading to the wilting of the stem fibres which then kills the plant.* R. solanacearum *is believed to have originated in Brazil in the 1890s, and by the early part of the twentieth century was found to infect at least twenty native plant species like potatoes, tomatoes and chillies. As new crops were planted across the world, the bacterium spread with unimaginable speed as it found new hosts to infect. By the mid-1960s, it is estimated that* Ralstonia *could have infected more than 250 plant species, and by 2007 nearly 450 plant species globally were susceptible to its infection. With such a wide range of hosts, entire farms with multiple crops can be devastated within a short time.*

Therapy in Poland. Hopeless cases of multi- and extremely drug-resistant (MDR/XDR) bacterial infections from all over the world trickle into these facilities, and most, if not all, leave for home completely cured. These barebones, minimally endowed institutions, have quietly demonstrated that phage therapy can be implemented successfully, especially for gastrointestinal diseases and diarrhoeas, respiratory diseases, wounds and burns, all of which are commonly infected by highly drug-resistant bacteria. Phage therapy has so far been used to help individual sufferers with intractable infections that conventional antibiotics and other means have failed to cure. Such patients seek help from institutions like Eliava and Hirschfield, or from individual phage therapy practitioners. What is missing are proven regimens that can be used to cure common bacterial infections like cholera or infections in burns, wounds, hospital-acquired and post-operative infections. We will need to develop clear protocols for phage therapy, setting up certifying laboratories and other institutions to regulate clinical practices. It is relatively easier and more economical to find a specific phage that can curb an epidemic of fast-spreading bacterial disease than making a vaccine from scratch or finding new antibiotics. The real challenge would be to get the pharmaceutical industry and governments to take up such research. It will mean decentralizing and

This image (3) of a mixed plantation of papaya, tomatoes and oranges was taken in eastern Colombia in 2018 and shows widespread damage caused by R. solanacearum. *One specific strain that causes the brown rot of potatoes wipes out US$950 million worth of crops annually, and is listed on the US Agricultural Bioterrorism Act of 2002. The only known natural enemy of* R. solanacearum *is a bacteriophage,* Ralstonia *phage phiRSL1 (look closely for the white specks inside the bacterium in 1). Field trials to defeat this highly pathogenic bacterium using phage therapy are currently under way, and some have shown promising results.*

making medicine more 'local'. In fact, if we take up phage therapy in medicine and personalised health, its application will be like the internet of things—there will be so many aspects to it that no single company or expert will be able to capture or comprehend it in all its dimensions. Phage therapy will shake up the current one-size-fits-all approach to health and treatment and place greater emphasis on diagnosis, identifying the right phage, creating repositories of phages and educating doctors and the public alike. This is not going to be easy, but given that we may be running out of viable options and ideas for treatment, phage therapy can become stand-alone or complementary to existing means of cure.

The alarming emergency of antibiotic resistance and the massive investments needed to find alternatives has now revived interest in phage therapy. But in the larger picture, concerted global efforts to find safer alternatives to antibiotics are few, far between and weak. In the US over the past two decades, the National Institutes of Health (NIH) and the CDC have shown limited commitment in terms of investments, but the approval of some recent projects suggests that there is hope of revival. The United States Army has funded research on phages to heal wound infections in Iraq war veterans, but this has not yet yielded large dividends for public health at large. Private companies like Nestlé have been exploring opportunities of using phages in nutraceuticals. In 2009, the British company Biocontrol Limited completed the first double-blind clinical trials which show that phage therapy is safe and effective for the treatment of chronic, antibiotic-resistant ear infections. As of October 2020, the Cochrane registry of clinical trials, clinicaltrials.gov and the WHO's ICTRP have listed 140 ongoing bacteriophage trials, which is an encouraging sign. But the efforts are spotty and a global consensus on the application of phage therapy—in human health, environmental mitigation and agriculture and livestock—is still missing.

Perhaps now more than ever before, we need to understand that our bodies are an ecosystem and each of us is a microcosm. Every pathogen that affects us has a set of its own pathogens—the enemy's enemy. This is how nature plays out. The revival of safe and responsive public health systems to address disease will depend upon how well we understand the ecology of viruses and the human microbiome in the context of disease and wellness. The harnessing of viruses to fight potential pathogens is a viable strategy for the advancement of medicine, both for public and individualised use—the future of human and planetary health may well lie in the hands (or rather, in the tails) of phages.

14
QUO VADIS?

Much as we may baulk at the idea, we are not unique. The human species, like all other living things, is an amalgam of creatures pieced together, gene by gene, and passed down by different life forms over deep time. Our genes were handed down to us from our ancestral ape, monkey, pig, shrew, gecko, fish, worm, grass, moss and bacterium, with several other creatures in between. Without genetic mutations, there would be no humans or, for that matter, any other life form that we see around us. These mutations—tiny errors in replicating the genetic code—occur randomly each time a cell (or virus) makes copies of itself, thereby becoming the starting point for an unexpected evolutionary journey. A very tiny number of mutations successfully create variation in a population. Natural selection then amplifies traits and creates variants which eventually can evolve into distinct species. These changes could be anything that confers an ability to thrive better in an evolving environment—blending a chameleon more effectively into a forest that is drying up; extending the necks of okapis and gerenuks so they can more easily nibble on the overhanging foliage of tall trees; or simply helping microbes evade a strong immune response and allowing them to attach themselves to a cell.

Viruses together with other microbes speed up the gene exchange between similar and often unrelated life forms. From a virus's perspective, all living creatures are genetic Frankensteins that they have stitched together. Nearly 8 per cent of our genes, for instance, are uniquely viral in origin, and despite thousands and millions of years of adding, editing and deleting genes, we remain genetically similar in many ways. A virus is capable of crossing over into us to cause infection because some of our shared genes, proteins and common cellular components enable it to establish itself.

Take SARS-CoV-2, one of our most recent chance encounters. A few generations ago, a population of viruses acquired changes in their genes (and proteins) from an animal host which provided them with the ability to infect humans. The virus binds to a protein receptor, ACE2 (or Angiotensin-converting enzyme 2; angiotensin is a protein hormone that causes blood vessels to become narrower and is vital for maintaining blood pressure and fluid balance in the body), which it recognises in us. ACE2 is found in several mammals, and in us they are primarily present in our lungs and guts. When virologists teased open the RNA of SARS-CoV-2, they found that it carried genes that are found in a variety of animals—in other viruses, bacteria, slugs, small mammals, and perhaps unsurprisingly therefore, nearly 94 per cent of its genes are found in humans too. But a handful of its genes are completely unique to it and their functions are yet unknown. This makes it difficult to predict how SARS-CoV-2 will behave once it enters our bodies.

Our technological progress, our ability to acquire instant (albeit often ephemeral) gratification, has lulled us into believing that we possess enough power to subdue and manipulate nature. We choose which relationships we want to foster and which we will cull and sever, and try to make nature serve us selectively and indefinitely. This kind of brinkmanship makes us feel that we can control nature, but as we have been slow to realise, this control is delusional. It is time we stopped looking at nature as a pliable variable or as an entity that impedes 'progress', or a tool of one-upmanship. I say this now because there are controversies about the origin of SARS-CoV-2—did the virus jump from an animal to unsuspecting buyers in a wet market or did it leak from a lab? Was it designed as a bioweapon? For me, these questions are relevant only if we have the courage to take corrective action and hold institutions or governments accountable. Otherwise this is a futile blame game best left to politicians and diplomats. The origin of the virus could be significant for science if it helps us determine the virus's lineage and identify its potential hosts which will help us plan future strategies.

Regardless of how it originated, SARS-CoV-2 reminds us that provoking nature beyond a point can lead to unimagined and irreversible consequences. Anyone who understands nature's processes knows that soils, mud, detritus, mulch, sand, gravel, grit and rock are crucial pieces in the climate change story, as are the ocean currents, wind circulation, the shapes and size of land masses and, of course, life forms—especially microbes, the principal primary energy producers on Earth that regulate the bulk of Earth's carbon-oxygen cycle. When all these small pieces come together, they power Earth's engines. This engine is an enormous, planetary-scale, biogeochemical reactor—but it starts from small things. Viruses cause a billion infections a second (10^{29} infections a day). They tinker and shuffle genes at great speed, creating possibilities of making new varieties of life. Like geological processes which create and shape diverse landscapes, viruses and microbes enable speciation and finds ways to fit new entities within ecosystems that their predecessors have shaped. Each ecosystem—a tropical forest, a vast grassland, a small pool or even microbiomes within every individual creature—has been thousands and even millions of years in the making, and is constantly evolving. These unceasing interactions of life forms in an ecosystem rest on the bedrock of microbial and mineral activity. Microbes are not only engines of evolution but they also digest, produce, process, ferment, break down, recycle, reformulate and synthesise chemicals faster and more efficiently than any human-made machine.

Every microbiome is distinct and is linked with our immediate environment. The microbiomes inside us are thriving and virtually every ounce of all our body fluids, tissues and organs of our bodies have distinct viruses that usually do more good to us than harm. Even parts of our bodies that were once considered sterile—our brain and spinal fluids, heart or kidneys—harbour specific viruses within them. If any ecosystem suffers injury or repeated insults, it cannot be restored to its original state even with the most advanced scientific effort.

Like any other pathogen that established itself as a persistent disease, SARS-CoV-2, too, was a chance occurrence. We enabled its crossover through destruction of habitats and trade in wildlife. And once the outbreak occurred, mass movements of people, weak and bigoted science, the fragmented response by agencies, a blunderbuss of regulations and distrust between states, sustained the spread and evolution of the novel virus. Using a global database of the virus's RNA collected from across the world, virologists have been able to monitor small mutations of the virus's 29,903-letter RNA code (our DNA has nearly 3 billion, gut-dweller

E. coli genome has 4.6 million). Two COVID-causing viruses isolated from patients anywhere in the world differ by an average of just ten RNA letters out of 29,903. With such slight changes, the virus becomes either more infective or more transmissible and several variants also peter out into oblivion. A couple of mutations makes a variant, and it takes an accumulation of several mutations to create a distinct species. This usually occurs when the virus passes through several different populations and encounters new immunological and other challenges (drugs used during treatment, for example). Variants of SARS-CoV-2 may be a long way away from becoming distinct species. Or not, since dogs, cats, tigers, minks and other animals are getting infected by interacting with humans. What we do know is that as it sets off on its uncharted journey, SARS-CoV-2 is opening up possibilities for other microbes to invade and colonise our microbiomes, and thus, change their fate, as well as ours. For SARS-CoV-2, the pandemic is not just a one-off chance but an evolutionary moment. The effects of the pandemic will not wear off any time soon. The virus has triggered massive changes starting with our bodies and embracing the body politic and these will probably stay with us for a very long time. Perhaps forever.

The pandemic has once again highlighted how little we know about the microbial world and how difficult it is for human technology to keep up with it. There are many new viruses being recorded each year. A single paper published in 2019 by an international team of scientists that surveyed 145 marine sites across the world, identified 1,95,728 *new* marine DNA viruses and entities. Since 2014, at least one new virus has been added to the list of known viruses that reside in our lungs, and in the last count, at least twenty of them have been found living in just the respiratory tract.

We cannot outnumber or even outmatch viruses. We must find ways to make peace with microbes (and viruses, in particular). Since we are living in a time when we are trying to set right past misdemeanours and political gaffes, we perhaps also need to begin building some appreciation for the microbe-kind and recognise their contributions. It comes as a surprise to me that even the most erudite scientists, popular science writers, broadcasters, doctors and public health experts, casually use labels like 'bug', 'critter', 'pest', 'intruder', among others, for microbes and the minutiae. Perhaps we should begin by according microbes the right honorific. If we continue to speak of microbes and viruses using the lexicon of war, then it is a war we have already lost. It is ironic that we are supposedly at war with this invisible force, which can cause disruption, but which also binds all life.

Viruses are a part of us. We cannot eliminate them without making ourselves extinct. We must begin to abandon our unending campaigns and eradication programmes that deploy toxic technologies to control diseases. We need sanity on how we use antibiotics, synthetic chemicals and disinfectants, and we need to stop 'over-medicalizing' our lives. As we stare at the precipice of an uncertain future and face multiple threats, we need to find solutions to these unsurmountable challenges we have created for ourselves. We can begin by understanding the many moving pieces that makes nature tick and find nature's way of healing itself. Can we use viruses and microbes to manipulate photosynthesis to reverse climate change on a global scale? Can we work with microbes to climate-proof food security? Will we leverage microbes to free ourselves from the use of toxic antibiotics and pesticides? Can we comprehend the mind-boggling symmetries and durable shapes, diverse processes and inner working of viruses to inspire the design of efficient and safer nanotechnology? We have only just begun to understand and harness the goodness and versatility of viruses and microbes to advance chronic disease management, cancer control and immunotherapy. There is so much more we can do if only we understood them better.

What inspiration can we take from this moment? This past year, nature has been telling us that we must slow down our greed machine. We need an anthropo-pause. We must awaken to the fact that we can share our spaces with nature in its many forms. The old ways were not 'normal' in any way, and that is not the normal that we should strive to return to. If we are to survive in the future as a civilization and as cultures, now is our moment to find ways to share our planet with all the life on Earth. And this change should start with our smallest denizens. It may not seem so at the moment but viruses are our friends.

NOTES

What does not feature in the main text is in this excursus. While I may attempt to present evidence and information objectively or try covering it in a 'positive' light, certain neutral and negative biases may have crept in. The title and subject of the book itself may be construed as being one. I have provided short descriptions to each reference where possible. For brevity's sake, and much to the dismay of the scientifically inclined, I have used MLA citations.

In general, I relied on several textbooks to revisit fundamental concepts; these include: Flint, Jane S., Vincent R. Racaniello, Glenn F. Rall, Anna Marie Skalka, Lynn W. Enquist (eds) *Principles of Virology*, 4th Edition, ASM Press and Wiley, August 2015; Norkin, Leonard, *Virology: Molecular Biology and Pathogenesis*, ASM Press, 2009; and Roossinck, Marilyn J., *Virus: An Illustrated Guide to 101 Incredible Microbes*, Princeton University Press, 2017. The International Committee on Taxonomy of Viruses' (ICTV) website, available at https://talk.ictvonline.org/ (14 July 2021) proved to be an invaluable tool to understand complex ideas.

1. Bounty

That all the SARS-CoV-2 can fit into a Cola can, see https://theconversation.com/all-the-coronavirus-in-the-world-could-fit-inside-a-coke-can-with-plenty-of-room-to-spare-154226 (9 June 2021).

I have presented a broad-brush and cursory picture of the microbial world. To explain a little further, the eukaryotes are made up of all animals and plants. Other than all things visible to the naked eye like plants, and animals, the eukaryotes comprise microbes with hairlike organelles (called ciliates); fungi comprise yeasts, moulds and mushrooms and have a distinct compound called chitin in their cell walls, which incidentally makes insects crunchy; flagellates are unicellular organisms or colonists with one or more whiplike appendages called flagella which cause movement; and finally, microsporidia which is a group of spore-forming unicellular parasites that are closer to fungi than single-celled microbes. For a wonderful overview, see Tudge, Colin, *The Variety of Life: A Survey and a Celebration of All the Creatures that Have Ever Lived*, Oxford University Press, 2000.

For numbers of viruses and bacterial cells, see Suttle, C.A., 'Viruses in the Sea', *Nature* vol. 437 (2005): 356–61. For viruses far outnumbering other life forms and how the estimate of 10^{31} was reached, see Hendrix, R.W. et al., 'Evolutionary Relationships among Diverse Bacteriophages and Prophages: All the World's a Phage', *Proceedings of the National Academy of Sciences of the United States of America* vol. 96, no. 5 (1999): 2192–97. That bacteriophages can be stretched for 100 million light years, and that there are 100 million times as many bacteria in the oceans (1.3×10^{29}) as there are stars in the known universe, see Mushegian, A.R., 'Are There 10^{31} Virus Particles on Earth, or More, or Fewer?', *Journal of Bacteriology* vol. 202, no. 9 (9 April 2020).

For estimates made on viruses inhaled in each lungful, see Prussin, A.J., E.B. Garcia and L.C. Marr, 'Total Virus and Bacteria Concentrations in Indoor and Outdoor Air', *Environ Sci Technol Lett.* vol. 2, no. 4 (2015): 84–88; and Willner, Dana et al., 'Metagenomic Analysis of Respiratory Tract DNA Viral Communities in Cystic Fibrosis and Non-cystic Fibrosis Individuals', *PLoS one* vol. 4, no. 10 (9 October 2009).

Microbes show the same geographic patterns of diversity, see Brown, James W., *Principles of Microbial Diversity*, ASM Press, 2014.

There are 100 million times as many bacteria in the oceans (1.3×10^{29}) as there are stars in the known Universe, see Suttle, Curtis A., 'Viruses: Unlocking the Greatest Biodiversity on Earth', *Genome* vol. 56, no. 10 (2013): 542–44.

The average number of microbes in a single teaspoon of soil (10^9) is the same as the human population of Africa and other astronomical numbers like microbes that live in the human jowl or dental plaque, see the editorial, 'Microbiology by Numbers', *Nature Reviews Microbiology* vol. 9, no. 9 (2011): 628.

About four-fifths of all animals are nematodes. See, van den Hoogen, Johan et al., 'A Global Database of Soil Nematode Abundance and Functional Group Composition', *Scientific Data* vol. 7, no. 1 (26 March 2020): 103.

On conditions for multicellular life to emerge, see Pennisi, E., 'The Momentous Transition to Multicellular Life May Not Have Been so Hard After All', *Science* (28 June 2018).

On at least 219 viruses capable of infecting humans, see Woolhouse, Mark et al., 'Human Viruses: Discovery and Emergence', *Philosophical Transactions of the Royal Society of London, Series B, Biological Sciences* vol. 367, no. 1604 (2012): 2864–71.

On modern biological taxonomy, see Watson, Mark F., Chris Lyal and Colin Pendry, *Descriptive Taxonomy: The Foundation of Biodiversity Research*, Cambridge University Press, 2015.

For one of the earliest discussions on why viruses don't have colour, see https://www.virology.ws/2009/02/23/what-color-is-a-virus/ (14 July 2021). I benefited from reading articles that appeared in *The New York Times,* available at https://www.nytimes.com/2020/04/01/health/coronavirus-illustration-cdc.html (14 July 2021); and the *Conversation,* available at https://theconversation.com/scary-red-or-icky-green-we-cant-say-what-colour-coronavirus-is-and-dressing-it-up-might-feed-fears-134380 (14 July 2021).

Carl Woese argued that molecular data could provide a practical metric for assessing evolutionary relationships between cultivated microorganisms in the context of 3.5 billion to 3.8 billion years of evolutionary history. He reasoned that the RNA component of the cell's protein synthesizing machinery evolved very slowly because it interacted with all other proteins in the cell either during their synthesis or as part of the ribosome, the nucleic acid-protein complex that assembles amino acids into proteins by a nucleic acid templating process. Because of its slow pace of evolution, similarities between ribosomal RNA (rRNA) sequences reflect the extent of genetic similarity between any two organisms. Closely related taxa have nearly identical rRNA sequences, while microbes that diverged from each other hundreds of millions of years ago exhibit greater levels of nucleotide variation. For the first time it became possible to infer objective dichotomous evolutionary branching patterns that describe taxonomic relationships for cultured microbes. For Carl Woese's classification and creating the label archaea, see David Quammen, *The Tangled Tree: A Radical New History of Life,* HarperCollins, UK, 2016, and the path-breaking paper by Woese and Fox which described archaea: Woese, C.R. and G.E. Fox, 'Phylogenetic Structure of the Prokaryotic Domain: The Primary Kingdoms', *Proceedings of the National Academy of Sciences of the United States of America* vol. 74, no. 11 (1977): 5088–90.

For all numbers, classifications and description of viruses, I relied on the extensive catalogues provided by the International Committee on the Taxonomy of Viruses (ICTV). In May 2020, ICTV changed its classification code to allow a fifteen-rank classification hierarchy which attempts to cover the entire spectrum of genetic divergence in the virosphere; see International Committee on Taxonomy of Viruses Executive Committee, 'The New Scope of Virus Taxonomy: Partitioning the Virosphere into 15 Hierarchical Ranks', *Nature Microbiology* vol. 5, no. 5 (May 2020): 668–74.

2. A Whole New World

On the treatises of Copernicus and Vesalius, and science and society in sixteenth-century Europe, see *The Invention of Science: A New History of the Scientific Revolution.* David Wootton, HarperCollins, New York, 2016, p. 778.

For an early history of optics, telescopes and microscopes, see Croft, William J., *Under the Microscope: A Brief History of Microscopy*, World Scientific Publishing Company, 2006; and Bradbury, S., *The Microscope: Past and Present*, Pergamon, 1968.

Pioneering studies using the microscope of Francesco Stelluti (1630) *Persio Appresso*; Giovanni Battista Hodierna (1644) *Opuscoli del dottor Don Gio.*; Jan Swammerdam (1659) *Bybel der natuure, door Jan Swammerdam-Amsteldammer-Of Historie der insecten*; Sir Robert Hooke (1665) *Micrographia*; Sir William Ramesey (1668) *Helmintholgia*; Francisci Redi (1671) *Patritii Aretini Experimenta circa generationem insectorum*; Nehemiah Grew (1682) *The Anatomy of Plants*; Marcello Malpighi (1673) *De Formatione de pulli in ovo*; Louis Joblot (1718) *Descriptions et Usages de Plusieurs Nouveaux Microscopes tant Simples que Composez*; Antonie van Leeuwenhoek (1722) *Arcana*, are available at https://www.biodiversitylibrary.org/ (14 July 2021).

For biographies of Hooke and van Leeuwenhoek, I referred to Egerton, Frank N., 'A History of the Ecological Sciences, Part 16: Robert Hooke and the Royal Society of London', *Bulletin of the Ecological Society of America*, vol. 86, no. 2 (2005): 93–101; and Ruestow, Edward G., *Physics at Seventeenth and Eighteenth-Century Leiden: Philosophy and the New Science in the University, Archives Internationales D'histoire des Idees/International Archives of the History of Ideas book series* (ARMI, vol. 11), Leiden, 1973, respectively.

The Reichenbach affair is recounted in Klemm, Friedrich, *A History of Western Technology*, London: George Allen and Unwin Ltd, 1959, pp. 259–61.

English astronomers like Frederick William Herschel, John Goodricke and Thomas Hutchins, and microscopists had adopted the use of German lenses. Deutsches Museum, Munich Handschriften. Des DM. 8277; HS 6168: Tagebuch von Georg von Reichenbach. The episode is recounted in Friedrich Klemm, *A History of Western Technology*, London: George Allen and Unwin Ltd, 1959, pp. 259–61.

On the early German microscope trade and industry and its rise to dominance, see Jackson, Myles W., *Spectrum of Belief: Joseph von Fraunhofer and the Craft of Precision Optics*, MIT Press, 2000; and Audrieth, L.F. and H.I. Chinn, The Organization of Science in Germany (International science reports, no. 2.), National Science Foundation and the Office of International Science Activities, Washington DC, 1963.

For the discovery of the action of a virus as opposed to other pathogens by Adolf Mayer, Dmitri Ivanovsky and Martinus Willem Beijerinck, see Calisher, Charles H. and M.C. Horzinek (eds), *100 Years of Virology: The Birth and Growth of a Discipline*, 1999, Springer Books.

We now know that some bacteria produce toxins that rapidly spread to other organs or even the entire body of the host. Such toxins (called exotoxins, exo-out) are among the most potent, unit for unit, of all toxic substances. In the 1890s, bacteriologists discovered that bacterial exotoxins spread to other organs quickly but found no bacteria but only traces of the toxin. There is no clear reason as to why some bacteria would spend their energy to poison other organs and yet multiply in the preferred site, given that they have successfully evaded the host's immunity and have commenced to multiply without the threat of being killed. The tetanus bacilli, for example, produces a toxin which diffuses quickly through the blood and causes muscles and nerves to fail, and if left untreated, causes death. Beijerinck thought that the disease he observed in tobacco could have been a bacterial (exo-)toxin.

The root of the word is derived from Sanskrit (*vish*), and in its Latinised meaning, denoted venom or poison, see Zimmer, Carl, *A Planet of Viruses* (Second Edition), University of Chicago Press, 2015.

On the July 1885 Joseph Meister episode and birth of the rabies vaccine, see Geison, Gerald L., *The Private Science of Louis Pasteur*, Princeton University Press, 1995.

For W.M. Stanley's experiments with TMV and its crystallization properties, see Hawkes, Peter W., *Advances in Imaging and Electron Physics*, Academic Press, 2014; and Rifkind, David and Geraldine Freeman, *The Nobel Prize-Winning Discoveries in Infectious Diseases*, Academic Press, 2005. On the controversy of TMV, see Wilkinson, L., 'The Development of the Virus Concept as Reflected in Corpora of Studies on Individual Pathogens: Lessons of the Plant Viruses—Tobacco Mosaic Virus', *Medical History* vol. 20, no. 2 (1976): 111–34.

For 'Key Events in the History of Electron Microscopy', see Haguenau, F., P.W. Hawkes, J.L. Hutchison, B. Satiat-Jeunemaître, G.T. Simon et al., *Microscopy and Microanalysis*, Cambridge University Press, vol. 9, no. 2 (April 2003): 96–138.

Among the earliest that they published was a wing of a housefly (1935), and they recalibrated their assembly of equipment and saw small things like diatoms, bacilli and a single iron filing (1937). See Ruska, Ernst, 'The Development of the Electron Microscope and of Electron Microscopy', Nobel Lecture, 8 December 1986, available at https://www.nobelprize.org/uploads/2018/06/ruska-lecture.pdf (14 July 2021).

The Berlin Group reported that when tobacco mosaic virus was put under the electron microscope, Ruska and team saw slinky rods arranged neatly in bundles, or 'cigarette-shaped particles'. See Kausche, G.A., E. Pfankuch and H. Ruska, 'Die Sichtbarmachung von pflanzlichem Virus im Übermikroskop', *Naturwissenschaften* vol. 27, no. 18 (1939): 292–99.

For the microscope as an invaluable tool in the fight against unknown diseases and useful in events like bioterror, see Goldsmith, Cynthia S. and Sara E. Miller, 'Modern Uses of Electron Microscopy for Detection of Viruses', *Clinical Microbiology Reviews* vol. 22, no. 4 (2009): 552–63.

For the first account of bacteriophage action viewed under the electron microscope by the Berlin Group, see Ruska, H., 'Die Sichtbarmachung der bakteriophagen Lyse im Übermikroskop', *Naturwissenschaften*, vol. 28, no. 3 (1940): 45–46.

3. Supersize Me

For evolution of early life, I recommend Kirschvink, Joe and Peter Ward, *A New History of Life: The Radical New Discoveries about the Origins and Evolution of Life on Earth*, Bloomsbury, 7 April 2015; and for macroevolution and diversification, see Zimmer, Carl, *At the Water's Edge: Fish with Fingers, Whales with Legs and How Life Came Ashore but Then Went Back to Sea*, Atria Books, August 2014.

On the 1992 discovery by Timothy Rowbotham, and the identification of a giant virus by Didier Raoult and Bernard La Scola, see Colson, P. et al., 'Giant Viruses of Amoebae: A Journey through Innovative Research and Paradigm Changes', *Annual Review of Virology* vol. 4, no. 1 (2017): 61–85.

For the Gram stain process, see Schmidt, Thomas M., *Encyclopaedia of Microbiology* vol. 1, Academic Press, 2019.

On the discovery of mimivirus reported in *Science*, see La Scola, B. et al., 'A Giant Virus in Amoebae', *Science*, vol. 299, no. 5615 (2003): 2033; and Colson, P. et al., 'Giant Mimiviruses Escape Many Canonical Criteria of the Virus Definition', *Clinical Microbiology and Infection: The Official Publication of the European Society of Clinical Microbiology and Infectious Diseases* vol. 25, no. 2 (2019): 147–54.

On the first description of medusavirus from a hot spring in Japan, see Yoshikawa, G. et al., 'Medusavirus, a Novel Large DNA Virus Discovered from Hot Spring Water', *Journal of Virology* vol. 93, no. 8 (3 April 2019).

On other giant viruses, pithovirus, see Levasseur, Anthony et al., 'Comparison of a Modern and Fossil Pithovirus Reveals Its Genetic Conservation and Evolution', *Genome Biology and Evolution* vol. 8, no. 8 (25 August 2016): 2333–39.

Mimiviruses that live within amoebae have DNA comparable in size to that of bacteria and in addition carry numerous cellular enzymes, see Raoult, D., 'The Journey from Rickettsia to Mimivirus', *ASM News*, vol. 71 (2005): 278–84.

Some, like Mimi, are so large that other smaller viruses (called virophages) can parasitise them. This was first described by La Scola, B. et al., 'The Virophage as a Unique Parasite of the Giant Mimivirus', *Nature* vol. 455, no. 7209 (2008): 100–04. In 2019, scientists reported the largest genome known in the virus world from a virus they labelled Choanovirus. The genome is so large that it codes its own protein including light-processing receptors called rhodopsins which are in many living organisms, including the retina of our eyes. The discovery of Choanovirus challenges the notion that viruses should be considered to be living in a state of suspended animation. See Needham, David M. et al., 'A Distinct Lineage of Giant Viruses Brings a Rhodopsin Photosystem to Unicellular Marine Predators',

Proceedings of the National Academy of Sciences of the United States of America vol. 116, no. 41 (2019): 20574–83.

For the anecdote of the discovery of *Cafeteria roenbergensis* virus by Tom Fenchel, see Fenchel, T., 'Marine Plankton Food Chains', *Annual Review of Ecology and Systematics* vol. 19, no. 1 (1988): 19–38.

On the vexing problem of the origin of viruses and the four proposed theories, see Flint, Jane S. et al. (eds), *Principles of Virology, 4th Edition*, ASM Press, August 2015; and Norkin, L., *Virology: Molecular Biology and Pathogenesis*, ASM Press, 2009.

On how medusavirus and amoebae may have traded genes on multiple occasions in the deep past and its possible reductive evolution, see Yoshikawa et al., op. cit.

On viruses reaching an evolutionary dead end, see Grubaugh, Nathan D. et al., 'We Shouldn't Worry When a Virus Mutates During Disease Outbreaks', *Nature Microbiology* vol. 5, no. 4 (2020): 529–30.

Among these, the most perplexing has been on the origin of HIV. In October 2000, a British medical journalist (Hooper, E., *The River: A Journey to the Source of HIV and AIDS*, Boston, MA, Back Bay Books, 2000) proposed a controversial hypothesis that HIV infection in humans originated by inadvertently testing oral polio vaccines in primates and apes in the Congo and other parts of Africa. It also stated that lab animals then transmitted the virus to wild animals and thus monkeys acquired the simian immunodeficiency virus (SIV), and also lions and other animals got their own forms of HIV. The book sparked legal action, laboratory investigations, several newspaper articles, and a spate of meetings in the highest scientific institutions. As soon as the book came out, rigorous tests of the original vaccine stock were analysed and confirmed that it contains neither HIV-1 nor chimpanzee DNA. See Plotkin, S.A. and H. Koprowski, 'No Evidence to Link Polio Vaccine with HIV . . .', *Nature* vol. 407, no. 6807 (2000): 941. For a paper which dispels the links with the emergence of HIV from polio vaccine, see Blancou, P. et al., 'Polio Vaccine Samples Not Linked to AIDS', *Nature* vol. 410, no. 6832 (2001): 1045–46. Second, a 2004 study published in *Nature* found that the strain of SIV affecting chimpanzees in the area where Hooper claimed the vaccine had been prepared using chimpanzee cells was genetically distinct from HIV strains. This refuted Hooper's claims from yet another angle: even if SIV-infected chimpanzee cells from that area had been used to make the vaccine, they could not have been the source of HIV. It is now agreed that HIV-1 originated from a chimpanzee lentivirus which encountered multiple cross-species transmissions before it crossed over into humans in the late nineteenth century in a region between Gabon and the Congo. See Sharp, P.M. et al., 'Origins and Evolution of AIDS Viruses: Estimating the Time Scale', *Biochemical Society Transactions* vol. 28, no. 2 (2000): 275–82. In 2008, a study which used molecular dating technology on various strains of HIV found that it was likely that certain African communities were infected prior to 1940, much earlier than the polio vaccine trials, and infections were caused by an entirely different strain that originated in Cameroon. See Worobey, M., M. Gemmel et al., 'Direct Evidence of Extensive Diversity of HIV-1 in Kinshasa by 1960', *Nature* vol. 455, no. 7213 (2 October 2008): 661–64.

Coronaviruses also emerged rather recently, just 700 or so years ago (around 1200 CE), and the lineage that infects bats, pangolins and us arose in the mid-twentieth century. See Boni, Maciej F. et al., 'Evolutionary Origins of the SARS-CoV-2 Sarbecovirus Lineage Responsible for the COVID-19 Pandemic', *Nature Microbiology* vol. 5, no. 11 (2020): 1408–17.

4. Us in the Virus

The quote of Sir Peter Medawar is from Medawar, Peter Brian and J.S. Medawar, *Aristotle to Zoos: A Philosophical Dictionary of Biology*, Harvard University Press, 1983. The entire short para reads: 'Inasmuch as viruses are made known only by their causing disease or other pathological changes, the existence of benign viruses having no ill effects remains conjectural. No virus is _known_ to do good: it has been well said that a virus is "a piece of bad news wrapped up in protein".' Italics as per the original text.

For basics on mechanisms of viral invasions, immunity, causing infection and replication, see Flint, Jane S., Vincent R. Racaniello, Glenn F. Rall, Anna Marie Skalka and Lynn W. Enquist (eds), *Principles of Virology, 4th Edition*, ASM Press and Wiley, August 2015; and Norkin, L., *Virology: Molecular Biology and Pathogenesis*, ASM Press, 2009.

On the human genome project, see Gibbs, Richard A., 'The Human Genome Project Changed Everything', *Nature Reviews Genetics* vol. 21, no. 10 (2020): 575–76; and Green, E.D., J.D. Watson and F.S. Collins, 'Human Genome Project: Twenty-five Years of Big Biology', *Nature* vol. 256, no. 7571 (1 October 2015): 29–31.

Scientists have found that there are entire sequences of RNA viruses called 'retroviruses'. For a good overview on retroviruses, see Skalka, Anna Marie, *Discovering Retroviruses*, Harvard University Press, 2019.

On the earliest acquisition and merger of an endogenous retrovirus (ERV), some studies have shown that small segments of bornavirus genes offered protection from bornaviral disease. They perhaps were assisted by a retrovirus that converted their RNA genome into DNA. See Hyndman, Timothy H. et al., 'Divergent Bornaviruses from Australian Carpet Pythons with Neurological Disease Date the Origin of Extant Bornaviridae prior to the End-Cretaceous Extinction', *PLoS Pathogens* vol. 14, no. 2 (20 February 2018). Reptiles and birds, the descendants of dinosaurs, are still most affected by bornaviruses, although horses and sheep sometimes get infected and die of neurological conditions, perhaps as a result of transmission by a shrew which acts as a reservoir.

Bornaviral elements were perhaps assisted by retroviruses or 'retro' elements in the genome, although they are not as good as ERVs at 'helping' other RNAs to insert into the genome. See Belyi, Vladimir A. et al., 'Unexpected Inheritance: Multiple Integrations of Ancient Bornavirus and Ebolavirus/Marburgvirus Sequences in Vertebrate Genomes', *PLoS Pathogens* vol. 6, no. 7 (29 July 2010).

On the deep history of ERVs and their transfer and acquisition, see Hayward et al., who state that 'we find only one single life-history character correlated with total ERV abundance: internal fertilization', implicating the protective environment of the female reproductive tract as important for virus survival and infection, while also suggesting a facilitating role for the fertilization process in the acquisition of 'endogenous' viral elements (EVEs). Hayward, A. et al., 'Pan-Vertebrate Comparative Genomics Unmasks Retrovirus Macroevolution', *Proceedings of the National Academy of Sciences of the United States of America* vol. 112, no. 2 (2015): 464–69.

Just as some viruses mimic the host cellular components with their genes, the viral envelope gene in rare cases can also be co-opted to serve host functions, particularly pertaining to the interaction between cell membranes such as cell fusion. See Cossart, P. and A. Helenius, 'Endocytosis of Viruses and Bacteria', *Cold Spring Harbor Perspectives in Biology* vol. 6, no. 8 (1 August 2014).

Vertebrate genome sequencing has confirmed that many viral genome sequences end up in the chromosome of the host cell, either as viral fragments or as complete genomes. For this, see Kryukov, Kirill et al., 'Systematic Survey of Non-Retroviral Virus-like Elements in Eukaryotic Genomes', *Virus Research* vol. 262 (2019): 30–36.

On the versatility of the mammalian placenta, Chuong, Edward B., 'The Placenta Goes Viral: Retroviruses Control Gene Expression in Pregnancy', *PLoS Biology* vol. 16, no. 10 (9 October 2018); and Dunn-Fletcher, C.E., L.M. Muglia, M. Pavlicev, et al., 'Anthropoid Primate-Specific Retroviral Element THE1B controls Expression of CRH in Placenta and Alters Gestation Length', *PLoS Biology* vol. 16, no. 9 (19 September 2018). The earliest ancestors of mammals, the reptile-like mammals, died out when the dinosaurs began to rule the Earth but left behind the two classes of ancestors of the modern mammals—the montremes like the platypus, and metatheria like kangaroos and us. For this, see Kaneko-Ishino, Tomoko and Fumitoshi Ishino, 'Evolution of Viviparity in Mammals: What Genomic Imprinting Tells Us about Mammalian Placental Evolution', *Reproduction, Fertility, and Development* vol. 31, no. 7 (2019): 1219–27.

The ancestral form of the mammalian placenta emerged roughly 130 million years ago, and this is mentioned in Griffith, Oliver W. and Günter P. Wagner, 'The Placenta as a Model for Understanding the Origin and Evolution of Vertebrate Organs', *Nature, Ecology & Evolution* vol. 1, no. 4 (23 March 2017): 72.

For a good overview of the human placenta, see Power, Michael L. and Jay Schulkin, *The Evolution of the Human Placenta*, Johns Hopkins University Press, 2012.

On the acquisition of another ERV gene in a distant primate ancestor, see On foamy viruses driving mammalian evolution, see Katzourakis, A. et al., 'Discovery of Prosimian and Afrotherian Foamy Viruses and Potential Cross Species Transmissions amidst Stable and Ancient Mammalian Co-evolution', *Retrovirology* vol. 11, no. 6 (4 August 2014).

A particular set of genes that flank ERVs called long terminal repeat (LTR) come into play and promote production of tissue-specific proteins in which they reside. Understanding the complex mechanism and orchestration of ERVs and host genes is a work in progress. We are discovering that some ERV genes which have been passed down from ancient mammalian ancestors have bestowed us with an inbuilt and inheritable defence mechanism called innate immunity. For early discoveries, read Chuong, Edward B. et al., 'Regulatory Evolution of Innate Immunity through Co-option of Endogenous Retroviruses', *Science* vol. 351, no. 6277 (2016): 1083–87.

About 400 million years or so ago an ancestor of the coelacanth became infected by novel foamy viruses and this triggered the development of a muscular fin in coelacanths. The reference for this is Ruboyianes, R. and M. Worobey, 'Foamy-like Endogenous Retroviruses Are Extensive and Abundant in Teleosts', *Virus Evolution* vol. 2, no. 2 (30 October 2016). Not all fish have foamy viruses, and some recently evolved fish like cichlids which have been evolving in freshwater lakes of Africa over the past tens of thousands of years and continue to do so, are gradually acquiring endo-foamy-viruses.

On how our ancestors in Africa acquired HERV-K and especially how HERV-K HML-2 (HK2) got incorporated in primates 30 million years ago, and whose descendants, us humans, bear its consequences, see Karamitros, T. et al., 'Human Endogenous Retrovirus-K HML-2 Integration within *RASGRF2* Is Associated with Intravenous Drug Abuse and Modulates Transcription in a Cell-line Model', *Proceedings of the National Academy of Sciences of the USA* vol. 115, no. 41 (9 October 2018):10434–39.

Genetic studies using bones of Neanderthal men, women and children and early *Homo sapiens* people from over 2,00,000 years until 40,000 years ago suggest that interbreeding occurred in both directions. While Neanderthals evolved in Eurasia, early *Homo sapiens* people dispersed from Africa already by 1,80,000 years ago. However, the early interbreeding left no trace because those *Homo sapiens* populations died out. It was only later contact after 55,000 years ago which left a Neanderthal genetic component which diminished over time to no more than 2 per cent in most living people's genomes; this probably came from Neanderthal men having babies with *Homo sapiens* women. Some of us today may have marginally more Neanderthal in us than others. Native West, Central and South Africans have no trace of Neanderthal, though they may have ancestry from other older African *Homo* species. Some who left Africa like the Andaman and Torres islanders may also have had little contact with Neanderthals, but they do show traces of the Denisovan genes, yet another *Homo* species from Eurasia who were closely related to Neanderthals. I thank Dr Rebecca M. Wragg Sykes for providing me with clarifications. In addition, her book, *Kindred: Neanderthal Life, Love, Death and Art*, Bloomsbury, October 2020, is a wonderful reference on the history of our ancestors.

On the exchange of HPV between sapiens and the Neanderthals, see Pimenoff, Ville N. et al., 'Transmission between Archaic and Modern Human Ancestors during the Evolution of the Oncogenic Human Papillomavirus 16', *Molecular Biology and Evolution* vol. 34, no. 1 (2017): 4–19. For reciprocal transmission of disease between sapiens and the Neanderthals, see Houldcroft, Charlotte J. and Simon J. Underdown, 'Neanderthal Genomics Suggests a Pleistocene Time Frame for the First Epidemiologic Transition', *American Journal of Physical Anthropology* vol. 160, no. 3 (2016): 379–88.

It may be possible that like KoRV, HIV too will get incorporated into the human genome in the future. This difficult question is partly answered in Sarker, Nishat et al., 'Genetic Diversity of Koala Retrovirus *Env* Gene Subtypes: Insights into Northern and Southern Koala Populations', *Journal of General Virology* vol. 100, no. 9 (2019).

On evolution and crossover of the gibbon ape leukaemia virus (GALV), see Denner, Joachim, 'Transspecies Transmission of Gammaretroviruses and the Origin of the Gibbon Ape Leukaemia Virus (GaLV) and the Koala Retrovirus (KoRV)', *Viruses* vol. 8, no. 12 (20 December 2016): 336.

Endogenised ERV, the cervid endogenous retrovirus (CrERV) in mule deer went one step further and have recently endogenised, although we don't know whether they are infectious or not. For this, see Kamath, Pauline L. et al., 'The Population History of Endogenous Retroviruses in Mule Deer (Odocoileus Hemionus)', *Journal of Heredity* vol. 105, no. 2 (2014): 173–87; and Elleder, Daniel et al., 'Polymorphic Integrations of an Endogenous Gammaretrovirus in the Mule Deer Genome', *Journal of Virology* vol. 86, no. 5 (2012): 2787–96.

On the 2006 experiments on HERV-K reinsertion in DNAs of hamsters and cats, see Dewannieux, Marie et al., 'Identification of an Infectious Progenitor for the Multiple-copy HERV-K Human Endogenous Retroelements', *Genome Research* vol. 16, no. 12 (2006): 1548–56.

On how ERVs may have spurred evolution, see, Bedford, Trevor and Harmit S. Malik, 'Did a Single Amino Acid Change Make Ebola Virus More Virulent?', *Cell* vol. 167, no. 4 (2016): 892–94; Molaro, Antoine et al., 'Evolutionary Origins and Diversification of Testis-specific Short Histone H2A Variants in Mammals', *Genome Research* vol. 28, no. 4 (2018): 460–73.

On banana streak virus (BSV), see Tripathi, Jaindra N. et al., 'CRISPR/Cas9 Editing of Endogenous *Banana Streak Virus* in the B Genome of *Musa* Spp. Overcomes a Major Challenge in Banana Breeding', *Communications Biology* vol. 2, no. 46 (31 January 2019).

For the role and progression of HERV-H, see Gemmell, P., J. Hein and A. Katzourakis, 'Phylogenetic Analysis Reveals that ERVs "Die Young" but HERV-H Is Unusually Conserved', *PLoS Computational Biology*, vol. 12, no. 6 (13 June 2016).

On a possible mechanism of how an HIV infection can activate human HERV-K, see Contreras-Galindo, Rafael et al., 'HIV Infection Reveals Widespread Expansion of Novel Centromeric Human Endogenous Retroviruses', *Genome Research* vol. 23, no. 9 (2013): 1505–13.

In 2015, medical scientists at the University of Rome suppressed the overexpression of one particular HERV (HERV-H) in a twelve-year-old boy with ADHD using a drug called methylphenidate over a period of six months, and mitigated his condition. See D'Agati, Elisa et al., 'First Evidence of HERV-H Transcriptional Activity Reduction after Methylphenidate Treatment in a Young Boy with ADHD', *New Microbiologica* vol. 39, no. 3 (2016): 237–39.

Many DNA viruses are capable of attacking both human cellular and metabolic processes simultaneously during infections. RNA viruses preferentially affect human protein processes within individual cells, across cells (intracellular) or across tissues and organs (intercellular). For this see, Durmuş, S. and Kutlu Ö Ülgen, 'Comparative Interactomics for Virus-Human Protein-Protein Interactions: DNA Viruses versus RNA Viruses', *FEBS Open Bio* vol. 7, no. 1 (4 January 2017): 96–107.

On the five HERV families, see Magiorkinis, Gkikas et al., '"There and Back Again": Revisiting the Pathophysiological Roles of Human Endogenous Retroviruses in the Post-Genomic Era', *Philosophical Transactions of the Royal Society of London, Series B, Biological Sciences* vol. 368, no. 1626 (12 August 2013).

The Chinese government appears to be aggressively funding research and trials in xenotransplantation. For a list of Chinese government funding, see Zheng, H., T. Liu, T. Lei, L. Girani, Y. Wang and S. Deng, 'Promising Potentials of Tibetan Macaques in Xenotransplantation', *Xenotransplantation* vol. 26, no. 1 (January 2019). For the process on xenotransplantation, see Sykes, Megan and David H. Sachs, 'Transplanting Organs from Pigs to Humans', *Science Immunology* vol. 4, no. 41 (2019); and Grow, Edward J. et al., 'Intrinsic Retroviral Reactivation in Human Preimplantation Embryos and Pluripotent Cells', *Nature* vol. 522, no. 7555 (2015): 221–25, which surveys a range of possible mechanisms and interactions in human embryonic cells.

Robin Weiss first discovered an endogenous retrovirus in the 1960s, when he found the jungle fowl and chicken genome. He is quoted as saying, 'If Charles Darwin reappeared today, he might be surprised to learn that humans are descended from viruses as well as from apes.' Specter, Michael, 'Darwin's Surprise: Why Are Evolutionary Biologists Bringing Back Extinct Deadly Viruses?', *New Yorker* (3 December 2007) is much used to explain evolution to undergraduates.

5. A Deep Control

The earliest cyanobacteria exploded on the scene around 3.6 billion years ago. They collaborated with other cyanobacteria to produce free oxygen. Over the next 1.2 billion years, so much oxygen was produced that it saturated the atmosphere with oxygen, created the ozone layer, and created new compounds like oxides and complex long-chained sugars. For this, see Kaufman, Alan J. et al., 'Late Archean Biospheric Oxygenation and Atmospheric Evolution', *Science* vol. 317, no. 5846 (2007): 1900–03. Cyanobacteria continued to ceaselessly proliferate along the margins where seas met land until the oxygen caused the planet to freeze over around 2.5 billion years ago for more than 200 million years. This is elegantly described in Luo, Genming et al., 'Rapid Oxygenation of Earth's Atmosphere 2.33 Billion Years Ago', *Science Advances* vol. 2, no. 5 (13 May 2016).

Cyanobacteria, at first by themselves and later in partnership with other microbes, formed colonies, and this is wonderfully described in Visscher, Pieter T. and John F. Stolz, 'Microbial Mats as Bioreactors: Populations, Processes, and Products', *Palæogeography, Palæoclimatology, Palæoecology*, vol. 219, nos 1–2, (2005).

For an overview of stromatolites, presence of modern stromatolite and the role and presence of ancient bacteriophage, see Desnues, Christelle et al., 'Biodiversity and Biogeography of Phages in Modern Stromatolites and Thrombolites', *Nature* vol. 452, no. 7185 (2008): 340–43.

The dominant cyanobacteria of extant open seas are *Prochlorococcus*, the smallest free-living photosynthetic organisms (diameter 0.5–0.7 μm). Outside of the viruses, *Prochlorococcus* are the most abundant creatures in the world, an estimated 2.9×10^{27} cells (compared to nearly 10^{29} bacteria in oceans. See Whitman, W.B. et al., 'Prokaryotes: The Unseen Majority', *Proceedings of the National Academy of Sciences of the United States of America* vol. 95, no. 12 (1998): 6578–83.

SAR11 and *Prochlorococcus* cooperate in using the organic minerals, but need the support of a few other microbes like sulphur-producing bacteria to help them flourish. Becker, Jamie W. et al., 'Co-culture and Biogeography of Prochlorococcus and SAR11', *ISME Journal* vol. 13, no. 6 (2019): 1506–19.

The photosynthetic bacterial community produces nearly 70 per cent of all the Earth's oxygen and sinks more than 4 gigatonnes of carbon dioxide (a gigatonne is a billion metric tonnes, GtC) every year, see https://climate.nasa.gov/ (15 July 2021). On the calculations on how much oceans sink (9 GtC) while forests and vegetation on land absorb about 11 GtC, and the oxygen budgets, see Prentice, I.C. et al., *The Carbon Cycle and Atmospheric Carbon Dioxide*, available at https://www.ipcc.ch/site/assets/uploads/2018/02/TAR-03.pdf (12 June 2021).

All the breathable oxygen is produced by marine microbes, see https://ocean.si.edu/ocean-life/plankton/every-breath-you-take-thank-ocean and references therein (11 July 2021).

Bacteriophages are the most abundant biological entities which feed on the most abundant bacteria, see Clokie, Martha Rj et al., 'Phages in Nature', *Bacteriophage* vol. 1, no. 1 (2011): 31–45.

Cyanophage are estimated to be up to 1×10^{30}. See Brüssow, Harald and Roger W. Hendrix, 'Phage Genomics: Small Is Beautiful', *Cell* vol. 108, no. 1 (2002): 13–16; and later revised to 4×10^{30} in Suttle, Curtis A., 'Viruses in the Sea', *Nature* vol. 437, no. 7057 (2005): 356–61.

The rate of viral infection in the oceans stands at 10^{23} infections per second, and these infections kill 20–40 per cent of all bacterial cells in the oceans each day. See 'Microbiology by numbers', *Nature Reviews Microbiology* vol. 9, no. 9 (2011): 628.

Phage-infected bacteria absorb more carbon before they die and sink to the ocean floor, compared to those that died uninfected. See Danovaro, Roberto et al., 'Marine Viruses and Global Climate Change', *FEMS Microbiology Reviews* vol. 35, no. 6 (2011): 993–1034.

The relationship of carbon dioxide and methane with bacteria that digest them and viruses that control their populations will be key for mitigating greenhouse gases. For this see Klasek, Scott et al., 'Microbial Communities from Arctic Marine Sediments Respond Slowly to Methane Addition during Ex Situ Enrichments', *Environmental Microbiology* vol. 22, no. 5 (2020): 1829–46; and the role of methane-digesting microbial communities and viruses in regulating and releasing methane, see Thamdrup, B, et al., 'Anaerobic Methane Oxidation Is an Important Sink for Methane in the Ocean's Largest Oxygen Minimum Zone', *Limnology & Oceanography* vol. 64 (2019): 2569–85.

When the cyanobacteria and other microbes are killed by viruses, their cells burst and release ions of carbonate which quickly react with free calcium ions in the seawater to make calcium carbonate. See Zeebe, R. and D. Wolf-Gladrow, *CO2 in Seawater: Equilibrium, Kinetics, Isotopes*, San Diego, CA: Elsevier, Inc., 2013. These coagulate and along with dead remains of other plankton descend like minute snowflakes in deep dark water.

Viruses and microbes don't just control the metabolism of carbon in water and on land, but also regulate phosphate (the key for energy transfer in cells); nitrogen (integral to make proteins); and sulphur (essential for several enzymes), which together are the building blocks of life. See Roux, Simon et al., 'Ecogenomics and Potential Biogeochemical Impacts of Globally Abundant Ocean Viruses', *Nature* vol. 537, no. 7622 (2016): 689–93.

Phage-infected bacteria which are in the throes of death absorb more carbon before they die and sink to the ocean floor, as compared to those which died uninfected (Sullivan, M.B., Ohio State University). The outer layers of some other free-floating organisms (called plankton) have carbon and silica-rich shells and skeletons which add to the sinking organic mass. These accumulate over tens and hundreds of millions of years, forming layers that are several hundred metres thick under the sea. See Read, Betsy A. et al., 'Pan Genome of the Phytoplankton Emiliania Underpins Its Global Distribution', *Nature* vol. 499, no. 7457 (2013): 209–13.

On the role of viruses and microbes in geological processes, see Pacton, Muriel et al., 'Viruses as New Agents of Organomineralization in the Geological Record', *Nature Communications* vol. 5, no. 4298 (3 July 2014).

On how viruses indirectly influence sulphur in marine food chains which impact precipitation, see Gao, Cherry et al., 'Single-Cell Bacterial Transcription Measurements Reveal the Importance of Dimethylsulfoniopropionate (DMSP) Hotspots in Ocean Sulfur Cycling', *Nature Communications* vol. 11, no. 1942 (23 April 2020).

Viruses adopt a slightly different process in shallow lakes and limestone caves. In shallow lakes of Central Asia and salt pans of Spain that rest over limestone terrane, biogeochemists (scientists who study the biological and chemical acts on geological features) and virologists have found that some specific viruses speed up carbon fixing by bacteria and help trap carbon dioxide within calcium and magnesium carbonate nodules. For this, see Zhu, Tingting and Maria Dittrich, 'Carbonate Precipitation through Microbial Activities in Natural Environment, and Their Potential in Biotechnology: A Review', *Frontiers in Bioengineering and Biotechnology* vol. 4, no. 4 (20 January 2016).

Viruses are pretty handy in reducing and preventing the outbreak of large floating masses of algae called algal 'bloom' which can spread over thousands of kilometres for several weeks and months. See these three reviews: Lehahn, Yoav et al., 'Decoupling Physical from Biological Processes to Assess the Impact of Viruses on a Mesoscale Algal Bloom', *Current Biology* vol. 24, no. 17 (2014): 2041–46; Frada, Miguel José et al., 'Zooplankton May Serve as Transmission Vectors for Viruses Infecting Algal Blooms in the Ocean', *Current Biology* vol. 24, no. 21 (2014): 2592–97; and Malits, A. et al., 'Enhanced Viral Production and Virus Mediated Mortality of Bacterioplankton in a Natural Iron-Fertilised Bloom Event above the Kerguelen Plateau', *Biogeosciences* vol. 11 (2014): 6841–53; Zimmerman, Amy E. et al., 'Metabolic and Biogeochemical Consequences of Viral Infection in Aquatic Ecosystems', *Nature Reviews Microbiology* vol. 18, no. 1 (2020): 21–34.

On the relationship of dominant cyanobacteria and the sulphur cycle, see Becker, Jamie W. et al., 'Co-culture and Biogeography of Prochlorococcus and SAR11', *ISME Journal* vol. 13, no. 6 (2019): 1506–19.

On the phosphorous cycle, see Pourtois, Julie et al., 'Impact of Lytic Phages on Phosphorus- vs. Nitrogen-Limited Marine Microbes', *Frontiers in Microbiology* vol. 11, no. 221 (21 February 2020).

6. Invaders, Hitch-hikers, Sentinels, Killers

Herpesviruses are large-sized viruses, with such large double-stranded DNA that the process of their integration with the host or even with bacterial genomes is very complex. There are at least eight common herpesviruses that infect us, and 90 per cent of the world's population carries at least one herpesvirus. See ICTV 9th report, available at https://talk.ictvonline.org/ictv-reports/ictv_9th_report/dsdna-viruses-2011/w/dsdna_viruses/91/herpesviridae (21 June 2021).

The origin of the herpesvirus family: ancestral herpesviruses emerged 400 million years ago, and infect bivalves, fish, frogs, reptiles, birds, mammals and us. See Brito, Anderson F. and John W. Pinney, 'The Evolution of Protein Domain Repertoires: Shedding Light on the Origins of the *Herpesviridae* family', *Virus Evolution* vol. 6, no. 1 (5 February 2020).

On the evolution and divergence of HSV1 and 2, see Wertheim, Joel O. et al., 'Evolutionary Origins of Human Herpes Simplex Viruses 1 and 2', *Molecular Biology and Evolution* vol. 31, no. 9 (2014): 2356–64.

On Divergence of HSV 2 and ChHV1, see Underdown, Simon J. et al., 'Network Analysis of the Hominin Origin of Herpes Simplex Virus 2 from Fossil Data', *Virus Evolution* vol. 3, no. 2 (1 October 2017).

Four of the six clades of HSV1 occur in East Africa, one each in East Asia and the others in Europe and North America. The virus left Africa with humans and reached Europe around 60,000 years ago. The evolution of the East Asian HSV-1 isolates is more complex because of several waves of migration that led to the peopling of East Asia. On why HSV2 is made up of just two lineages, see Casto, Amanda M. et al., 'Large, Stable, Contemporary Interspecies Recombination Events in Circulating Human Herpes Simplex Viruses', *Journal of Infectious Diseases* vol. 221, no. 8 (2020): 1271–79.

Studies on dentition, isotope analysis of dental enamel, microscopic scratches and shear marks of the teeth suggested that *P. boisei* preferred chewing shoots and leaves, and an occasional small animal if it could chance upon it. For this, see Grine, Frederick E. et al., 'Dental Microwear and Stable Isotopes Inform the Palæecology of Extinct Hominins', *American Journal of Physical Anthropology* vol. 148, no. 2 (2012): 285–317.

On recent HSV1 and HSV2 migration out of Africa, see Forni, Diego et al., 'Recent Out-of-Africa Migration of Human Herpes Simplex Viruses', *Molecular Biology and Evolution* vol. 37, no. 5 (2020): 1259–71.

For an excellent overview of the JCV and its spread, see Stoner, G.L. et al., 'JC Virus as a Marker of Human Migration to the Americas', *Microbes and Infection* vol. 2, no. 15 (2000): 1905–11. More recent dating using JCV can be seen, for example, in the Bering Strait crossing from Asia in to America. See Kitchen, Andrew et al., 'Utility of DNA Viruses for Studying Human Host History: Case Study of JC Virus', *Molecular Phylogenetics and Evolution* vol. 46, no. 2 (2008): 673–82.

For an overview on TTV, see Rezahosseini, Omid et al., 'Torque-Teno Virus Viral Load as a Potential Endogenous Marker of Immune Function in Solid Organ

Transplantation', *Transplantation Reviews, Orlando, Fla.* vol. 33, no. 3 (2019): 137–44.

On Ad36 and obesity, see Akheruzzaman, Md et al., 'Twenty-five Years of Research about Adipogenic Adenoviruses: A Systematic Review', *Obesity Reviews: An Official Journal of the International Association for the Study of Obesity* vol. 20, no. 4 (2019): 499–509.

For findings from the University of Wisconsin obesity experiment, see Atkinson, R.L. et al., 'Human Adenovirus-36 Is Associated with Increased Body Weight and Paradoxical Reduction of Serum Lipids', *International Journal of Obesity* vol. 29, no. 3 (2005): 281–86.

The term infectobesity has been coined by Professor Nikhil V. Dhurandhar, who proposed the adipogenic effect of the human adenovirus Ad36 on laboratory animals and also its association with human obesity.

Infection with adenovirus 36 (Ad36) produces obesity in animal models and is associated with risk of obesity within human populations, with children at particularly high risk. See Ponterio, Eleonora and Lucio Gnessi, 'Adenovirus 36 and Obesity: An Overview', *Viruses* vol. 7, no. 7 (8 July 2015): 3719–40.

Polyomaviruses, as their name suggests, also cause harm when they are activated. We currently estimate that both RNA and DNA viruses are responsible for nearly 15 per cent of all human cancers. For a more detailed survey of *polyomaviruses* and the evolutionary pathways, see Pena, Giselle P.A. et al., 'Human Polyomavirus KI, WU, BK, and JC in Healthy Volunteers', *European Journal of Clinical Microbiology & Infectious Diseases: Official Publication of the European Society of Clinical Microbiology* vol. 38, no. 1 (2019): 135–39; and Torres, Carolina, 'Evolution and Molecular Epidemiology of *Polyomaviruses*', *Infection, Genetics and Evolution: Journal of Molecular Epidemiology and Evolutionary Genetics in Infectious Diseases* vol. 79 (2020).

Recently, studies found that HIV-positive people who are co-infected with a common, non-pathogenic human flavivirus (labelled GB virus type C [GBV-C]) survive significantly longer than do HIV-positive individuals without GBV-C infection. See Stapleton, Jack T., 'GB Virus Type C/Hepatitis G Virus', *Seminars in Liver Disease* vol. 23, no. 2 (2003): 137–48.

7. A Spotty History of the Speckled Monster

An excellent overview of the pre-eradication era is by Behbehani, A.M., 'The Smallpox Story: Life and Death of an Old Disease', *Microbiological Reviews* vol. 47, no. 4 (1983): 455–509.

On the origin of poxviruses, see Babkin, Igor V. and Irina N. Babkina, 'The Origin of the Variola Virus', *Viruses* vol. 7, no. 3 (10 March 2015): 1100–12.

Viruses and other microbes that co-evolved in now-extinct mammal species are threatening to re-emerge and may interact with us in entirely different ways.

The evolutionary leap from rodents to humans and camels occurred about 4000 years ago. There is also a counter hypothesis which suggests that camels carried the poxvirus with them when they left America and entered Asia by crossing the Bering Strait, and an encounter with new mammals may have enabled the crossover into rodents like gerbils, and later it crossed over to humans. See Babkin and Babkina, 2015, op. cit. Also see Fenner, F., D.A. Henderson, I. Arita et al., *Smallpox and Its Eradication*, Geneva, World Health Organization, 1988, p. 1460.

Gerbil taxonomy is mired in confusion. Many texts and databases use both gerbil and *tatera* (as they are called locally in India). The naked-soled is also called northern savannah gerbil, *Gerbilliscus kempi* (earlier *Tatera kempi*), see IUCN database, The Red List of Threatened Species, IUCN, available at www.iucnredlist.org (13 July 2021).

There are seventy-one species of poxviruses listed under ICTV database 2021, available at https://talk.ictvonline.org/ (13 July 2021).

The rise of smallpox with agriculture around 10,000 BCE in the Fertile Crescent is discussed in Barquet, N. and P. Domingo, 'Smallpox: The Triumph over the Most Terrible of the Ministers of Death', *Annals of Internal Medicine* vol. 127, no. 8, Pt 1 (1997): 635–42.

See McNeill, William, author of Plagues and Peoples. 'Little wonder, then, that the Indians accepted Christianity and submitted to Spanish control so meekly. God had shown Himself on their side, and each new outbreak of infectious disease imported from Europe, and soon from Africa as well, renewed the lesson.' See McNeill, J.W., *Plagues and Peoples*, Anchor Press/Doubleday, 1976.

Variola minor was also called alastrim in the early eighteenth century, a term given by Portuguese slave traders from the word alastrar, or to spread out. It is believed that even mild smallpox could cause widespread death in immunologically naive indigenous people. See Ghio, Andrew J., 'Particle Exposure and the Historical Loss of Native American Lives to Infections', *American Journal of Respiratory and Critical Care Medicine* vol. 195, no. 12 (2017): 1673.

The Commander-in-Chief for North America, Jeffrey Amherst, in 1793 wrote to the commander of a fort (Colonel Bouquet) that had suffered a smallpox outbreak: 'Could it not be contrived to send the smallpox among those disaffected tribes of Indians? We must on this occasion, use every stratagem in our power to reduce them.' The same thought had crossed the mind of one lieutenant of the garrison, who on his initiative distributed disease-ridden items, documented the deed, and even filed for official reimbursement to cover the costs of the blankets and handkerchief used! The fort commander reimbursed them for the 'sundries got to replace in kind those which were taken from people in the hospital to convey smallpox to the Indians'. See D'Errico, P., *Jeffery Amherst and Smallpox Blankets*, 2001, available at https://people.umass.edu/derrico/amherst/lord_jeff.html (13 July 2021).

On the origins of vaccination, see Boylston, Arthur, 'The Origins of Vaccination: Myths and Reality', *Journal of the Royal Society of Medicine* vol. 106, no. 9 (2013): 351–54.

On Jenner's early experiments with using cowpox, see Jenner, E., *An Enquiry into the Causes and Effects of the Variolae Vaccinae, Known by the Name of the Cow Pox*, London: Sampson Low, 1798; and *The Origin of the Vaccine Inoculation*, London: D.N. Shury, 1801; *Instructions for Vaccine Inoculation*, London: D.N. Shury, 1801; On the Varieties & Modifications of the Vaccine Pustule, occasioned by an Herpetic State of the Skin, mimeo, London.

Edward Jenner insisted that his friend and fellow physician, Richard Dunning be credited for coining the term vaccination in 1803. See Le Fanu, W.R., 'Edward Jenner', *Proceedings of the Royal Society of Medicine* vol. 66, no. 7 (1973): 664–68.

It is possible that the principle of variolation existed in India but not vaccination. Two accounts of Bengal region by Coult (1731) and Holwell (1767) describe the process of inoculation by Brahmins. See Holwell, J.Z., 'An Account of the Manner of Inoculating for the Smallpox in the East Indies, with some Observations on the Practice and Mode of Treating that Disease in those Parts', T. Becket & P.A. de Hondt, London, 1767, available at https://archive.org/details/accountofmannero00holw/page/n55/mode/1up (16 July 2021); and 'Operation of Inoculation of the Smallpox as performed in Bengall', from Ro. Coult, Ro. to Dr Oliver Coult in 'An Account of the Diseases of Bengall', Calcutta, 10 February 1731, reproduced in Dharampal *Collected Writings*, Volume I, *Indian Science and Technology in the Eighteenth Century*, Goa: Other India Press, 2000.

David Arnold points out that the vaccination campaign by the East India Company was the first way in which the British-Indian government impinged directly on the lives of ordinary Indians. See Arnold, D., *Colonizing the Body: State Medicine and Epidemic Disease in Nineteenth Century India*, Berkeley: University of California Press, 1993. Arnold concludes that variolation was widespread in Bengal, Assam, Bihar and Orissa. It was also common around Varanasi (Benares), in parts of Punjab, in Rajasthan, Sindh, Gujarat and in scattered parts of Maharashtra and central India. A few historical accounts also provide insights on the role of Brahmins in variolation; for example, see Whitelaw, Ainslie, 'Observations Respecting the Small-pox and Inoculation in Eastern Countries; With Some Account of the Introduction of Vaccination into India', *Trans. R. Asiatic Society, Great Britain and Ireland* vol. 2, nos 52–73 (1830): 63.

On the early history of the pox in the Indian subcontinent, see Corrêa, G., R.J. de Lima Felne, *Lendas Da India, Lisboa*: Academia Real Das Sciencias, 1864, p. 192.

On Francis Whyte Ellis and the 'pious fraud', see Dominik Wujastyk, '"A Pious Fraud": The Indian Claims for pre-Jennerian Smallpox Vaccination', in Meulenbeld, G.J. and D. Wujastyk (eds), *Studies on Indian Medical History*, Groningen: Egbert Forsten, 1987, pp. 131–67, on p. 151. Another review which surveys the introduction of vaccination in India is Brimnes, Niels, 'Variolation, Vaccination and Popular Resistance in Early Colonial South India', *Medical History* vol. 48, no. 2 (2004): 199–228.

For rates of vaccination from 1810 to 1905, see James, S.P., *Smallpox and Vaccination in British India*, Thacker, Spink & Co., 1909, available at https://www.rarebooksocietyofindia.org/book_archive/196174216674_10154377835721675.pdf (15 July 2021).

On the use of children as couriers of the inoculum, see Murdoch, Lydia, 'Carrying the Pox: The Use of Children and Ideals of Childhood in Early British and Imperial Campaigns against Smallpox', *Journal of Social History* vol. 48, no. 3 (2015): 511–35.

The milkmaid's myth was written by Jenner's friend and biographer, John Baron, who was nominated through an execution of the will and estate of Jenner to write his account, which eventually got published several years after his death. See Baron, John, *Life of Edward Jenner, M.D., LL.D., F.R.S., with Illustrations of His Doctrine and Selections from His Correspondence* vols 1 and 2, Cambridge Library Collection, UK, pp. 1837-38.

For the observation on cuckoo behaviour, see the Royal Society for Jenner's landmark nineteen-page ornithological publication, *Observations on the Natural History of the Cuckoo*. Among several observations on British birds and their calls, nesting behaviour and diets, Jenner reported on how newly hatched cuckoos eject the eggs or fledglings of its foster parents from the nest. Jenner observed that the eviction was done when a peculiar depression between the scapulae (shoulder blades) of the young cuckoo disappears about twelve days after its birth, available at https://royalsocietypublishing.org/doi/pdf/10.1098/rstl.1788.0016 (19 July 2021).

On the formalin jars at the National Museum in Prague, and data of the sample, see Pajer, Petr et al., 'Characterization of Two Historic Smallpox Specimens from a Czech Museum', *Viruses* vol. 9, no. 8 (27 July 2017): 200.

On isolation and dating of the smallpox genome from a Lithuanian church crypt, see Duggan, Ana T. et al., '17th-Century Variola Virus Reveals the Recent History of Smallpox', *Current Biology* vol. 26, no. 24 (2016): 3407–12.

For the study on Viking skeletons and the variola genome which pushed back the date of the Lithuanian crypt by about 1000 years, see Mühlemann, B. et al., 'Diverse Variola Virus (Smallpox) Strains Were Widespread in Northern Europe in the Viking Age', *Science* vol. 369, no. 6502 (2020).

For a recent analysis of the smallpox genome from lancets from the American civil war, see Duggan, Ana T. et al., 'The Origins and Genomic Diversity of American Civil War Era Smallpox Vaccine Strains', *Genome Biology* vol. 21, no.1 (20 July 2020): 175.

On the 2017 assembling of genes to create variola using horsepox by Canadian researchers for $1,00,000 using mail-order DNA, see Noyce, Ryan S. et al., 'Construction of an Infectious Horsepox Virus Vaccine from Chemically Synthesised DNA Fragments', *PloS one* vol. 13, no.1 (29 January 2018); and subsequent ethical issues raised in Inglesby, Tom, 'Horsepox and the Need for a New Norm, More Transparency, and Stronger Oversight for Experiments that Pose Pandemic Risks', *PLoS Pathogens* vol. 14, no. 10 (4 October 2018).

For the cocoliztli epidemic in Mexico's highland regions, see Vågene, Åshild, *Enteric Fever in Sixteenth-Century Mexico*, available at https://natureecoevocommunity.nature.com/posts/29300-enteric-fever-in-sixteenth-century-mexico (15 January 2018).

On the prevalence and emergence of cowpox in Brazil, see Abrahão, Jônatas Santos et al., 'Outbreak of Severe Zoonotic Vaccinia Virus Infection, South-eastern Brazil', *Emerging Infectious Diseases* vol. 21, no. 4 (2015): 695–98; and Georgia Vora, Neil M. et al., 'Human Infection with a Zoonotic Orthopoxvirus in the Country of Georgia', *New England Journal of Medicine* vol. 372, no. 13 (2015): 1223–30; and buffalopox, which has recurred in South Asia remains diagnosed as a fever of unknown origin in the region, see Singh, R.K. et al., 'Buffalopox: An Emerging and Re-emerging Zoonosis', *Animal Health Research Reviews* vol. 8, no. 1 (2007): 105–14.

Monkeypox is related to smallpox and originated in West Africa and the Congo Basin, but animal trade and human migration have spread this across the world. The trade of gerbils as pets is a possible flashpoint of the crossing over of an unknown poxvirus. Then there is camelpox, which is found wherever camels are found, naturally or introduced; and at least nine species of poxviruses about which we know very little. An Orthopoxvirus species labelled Alaskapox which causes skin lesions has appeared and disappeared in Alaska twice in the last five years. It is possible that the virus was transmitted through animal–human contact, but this novel virus's origin remains unknown. See Gigante, Crystal M. et al., 'Genome of Alaskapox Virus, A Novel Orthopoxvirus Isolated from Alaska', *Viruses* vol. 11, no. 8 (1 August 2019): 708.

Some scientists are cautious and are watching for the evolutionary transformation of a virulent form of a poxvirus, see Shchelkunova, G.A. and S.N. Shchelkunov, '40 Years without Smallpox', *Acta Naturae* vol. 9, no. 4 (2017): 4–12.

WHO's Director General Tedros Adhanom Ghebreyesus's statement on the fortieth anniversary observation, available at https://www.who.int/news-room/detail/13-12-2019-who-commemorates-the-40th-anniversary-of-smallpox-eradication (22 June 2021).

For the media report by the Wuhan Municipal Health Commission of an unknown 'viral pneumonia' of unknown cause in Wuhan, see https://www.who.int/news-room/detail/29-06-2020-covidtimeline (22 June 2021).

8. Gut Feeling

For a good overview on the gut, the microbiome and health, I recommend Yong, Ed, *I Contain Multitudes: The Microbes Within Us and a Grander View of Life*, HarperCollins, 2018; and Enders, Guilia, *Gut: The Inside Story of Our Body's Most Under-Rated Organ*, Scribe UK, 2015.

On the 10 trillion human cells and 100 trillion bacteria ratio, see Luckey, T.D., 'Introduction to Intestinal Microecology', *American Journal of Clinical Nutrition* vol. 25, no. 12 (1972): 1292–94.

The 1977 review by Savage is referenced over 1000 times in the literature, often in the context of the estimate for the vast overabundance of bacteria over human cells. See Savage, D.C., 'Microbial Ecology of the Gastrointestinal Tract', *Annual Review of Microbiology* vol. 31 (1977): 107–33.

For the 'Letter to the Editor' which appeared in the March 2014 edition of *Microbe* magazine, see Rosner, J.L., 'Ten Times More Microbial Cells than Body Cells in Humans?', *Microbe* vol. 9, no. 47 (2014).

On the origin of the terms microbiota and microbiome, see Prescott, Susan L., 'History of Medicine: Origin of the Term Microbiome and Why It Matters', *Human Microbiome Journal* vol. 4 (June 2017): 24–25.

For Fermi's back-of-the-envelope method of estimation, see Moore, Joyce L., 'Back-of-the- Envelope Problems', submitted by University of California Berkeley for US Naval Research Centre (Unclassified document), 1987, available at https://apps.dtic.mil/sti/pdfs/ADA184105.pdf (15 July 2021).

Bacteria have the most bulk; about 93 per cent of all microbes in terms of mass, was estimated by Sender, Ron et al., 'Revised Estimates for the Number of Human and Bacteria Cells in the Body', *PLoS Biology* vol. 14, no. 8 (19 August 2016). Human-associated fungi (about 0.1 per cent of the gut microbial mass) are thankfully outnumbered by the bacteria because we do not want them to take over since fungi (like *Aspergillus*) are aggressively invasive.

Human microbiota primarily has bacteriophage and perhaps outnumbers the bacterial population by at least tenfold and makes up 5.8 per cent of the microbiome by weight. See Shkoporov, Andrey N. et al., 'ΦCrAss001 Represents the Most Abundant Bacteriophage Family in the Human Gut and Infects Bacteroides Intestinalis', *Nature Communications* vol. 9, no. 1 (14 November 2018): 4781.

One study of human faeces from different individuals estimated that the gut contains an estimated 1200 viral genotypes. See Breitbart, Mya et al., 'Metagenomic Analyses of an Uncultured Viral Community from Human Feces', *Journal of Bacteriology* vol. 185, no. 20 (2003): 6220–23; and up to 90 per cent are from viruses which we don't know anything about. For this, see Aggarwala, Varun et al., 'Viral Communities of the Human Gut: Metagenomic Analysis of Composition and Dynamics', *Mobile DNA* vol. 8 no. 12 (3 October 2017); and

Shkoporov, Andrey N., and Colin Hill, 'Bacteriophages of the Human Gut: The "Known Unknown" of the Microbiome', *Cell Host & Microbe* vol. 25, no. 2 (2019): 195–209. A more recent study, Camarillo-Guerrero, Luis F. et al., 'Massive Expansion of Human Gut Bacteriophage Diversity', *Cell* vol. 184, no. 4 (2021) has identified over 1,40,000 virus 'species' in the human gut, more than half of which have never been seen before. There are also tiny parasitic intestinal worms called helminths which are typically absent from people living in high-income nations but live in the gut of billions of people in low- and middle-income countries and affect them with varying degrees of severity.

Other shocks like fever, antibiotic use, surgery, extensive travel and encountering new cuisines and diets can upset your gut's microbial residents and therefore its stability, see David, Lawrence A. et al., 'Diet Rapidly and Reproducibly Alters the Human Gut Microbiome', *Nature* vol. 505, no. 7484 (2014): 559–63.

On the benefits of microbial transfer and its potential effect on autism, see Kang, Dae-Wook et al., 'Microbiota Transfer Therapy Alters Gut Ecosystem and Improves Gastrointestinal and Autism Symptoms: An Open-label Study', *Microbiome* vol. 5, no. 1 (23 January 2017): 10.

On links of the microbiome and how it can lead to obesity, see Ley, Ruth E. et al., 'Microbial Ecology: Human Gut Microbes Associated with Obesity', *Nature* vol. 444, no. 7122 (2006): 1022–23.

Some scientists regard the human gastrointestinal tract as 'one of the fiercest competitive ecological niches found in nature', see Klijn, Adrianne et al., 'Lessons from the Genomes of Bifidobacteria', *FEMS Microbiology Reviews* vol. 29, no. 3 (2005): 491–509.

On the colonization of microbes at birth and how it continues to alter and adapt throughout life, see Human Microbiome Project Consortium, 'Structure, Function and Diversity of the Healthy Human Microbiome', *Nature* vol. 486, no. 7402 (13 June 2012): 207–14; and on virome development, see Breitbart, Mya et al., 'Viral Diversity and Dynamics in an Infant Gut', *Research in Microbiology* vol. 159, no. 5 (2008): 367–73.

On a wide array of functions including immune system development, synthesis of vitamins and energy generation by microbes, see Qin, Junjie et al., 'A Human Gut Microbial Gene Catalogue Established by Metagenomic Sequencing', *Nature* vol. 464, no. 7285 (2010): 59–65.

On the role of resident phage and lung health, see Chang, Rachel Yoon Kyung et al., 'Phage Therapy for Respiratory Infections', *Advanced Drug Delivery Reviews* vol. 133 (2018): 76–86; and Waters, Elaine M. et al., 'Phage Therapy Is Highly Effective against Chronic Lung Infections with *Pseudomonas aeruginosa*', *Thorax* vol. 72, no. 7 (2017): 666–67.

The gut virome is also highly specific to each individual and dominated by bacteriophages, and this is described in Shkoporov, Andrey N. and Colin Hill, op. cit.

In infancy, bacteriophages and 'good' viruses are in even higher proportion which protect the young, and in a sense curate the gradual occupancy by 'good' bacteria. As good bacteria grow in the gut and other microbiomes, the bacteriophage population stabilises. See Beller, Leen and Jelle Matthijnssens, 'What Is (Not) Known about the Dynamics of the Human Gut Virome in Health and Disease', *Current Opinion in Virology* vol. 37 (2019): 52–57.

Each individual therefore develops a distinct microbiome, which changes with time. Even among twins, the composition of the microbiome may be similar for a short while but gradually changes to become very distinct. This is well described in Reyes, Alejandro et al., 'Viruses in the Faecal Microbiota of Monozygotic Twins and Their Mothers', *Nature* vol. 466, no. 7304 (2010): 334–38.

These patterns of diversity are driven by bacteriophages, thus the abundance and diversity of the microbiomes also reflects on the virome richness and diversity. For this, see Kernbauer, Elisabeth et al., 'An Enteric Virus Can Replace the Beneficial Function of Commensal Bacteria', *Nature* vol. 516, no. 7529 (2014): 94–98.

There is growing evidence which suggests that the composition of the intestinal microbiota which develops in infants and young children determines whether they will develop allergies or asthma as adults or not. See Zimmermann, Petra et al., 'Association between the Intestinal Microbiota and Allergic Sensitization, Eczema, and Asthma: A Systematic Review', *Journal of Allergy and Clinical Immunology* vol. 143, no. 2 (2019): 467–85.

Early results from experiments are showing promise for targeted and less toxic treatment of infections especially for those who develop frequent infection from *E. coli*. For this, read Dalmasso, Marion et al., 'Three New *Escherichia Coli* Phages from the Human Gut Show Promising Potential for Phage Therapy', *PLoS one* vol. 11, no. 6 (9 June 2016) and references therein.

On how to harness the microbiota in our pursuit of personalised medicine, see a good review in Mirzaei, Mohammadali Khan and Corinne F. Maurice, 'Ménage à Trois in the Human Gut: Interactions between Host, Bacteria and Phages', *Nature Reviews Microbiology* vol. 15, no. 7 (2017): 397–408.

A growing number of serious medical conditions that were formerly thought to be unlinked to microbes may actually have bacterial culprits. For instance, a natural gut resident called *Helicobacter pylori* can turn rogue under certain conditions and cause gastric ulcers which may lead to stomach ulcers. The actions of these bacteria lead to hardening of arteries and when disturbed by stress, these play a part in triggering heart attacks. Although a specific phage was discovered in 1989, its use to fight and remove the bacterium has not been explored. A good review is by Schmid, E.N. et al., 'Bacteriophages in Helicobacter (Campylobacter) Pylori', *Journal of Medical Microbiology* vol. 32, no. 2 (1990): 101–04.

A review of standard texts of gastroenterology like Yamada's *Textbook of Gastroenterology* and Harrison's *Gastroenterology and Hepatology*, and Murray;

Lange; Lippincotts; or Sherris Medical Microbiology maintained that there were a handful of species of good bacteria.

The first resident virus, a benign calcivirus, was reported in 1993, see Grohmann, G.S. et al., 'Enteric Viruses and Diarrhea in HIV-infected Patients. Enteric Opportunistic Infections Working Group', *New England Journal of Medicine* vol. 329, no. 1 (1993): 14–20.

For the April 2011 study from the European Molecular Biology Laboratory which labelled the four 'enterotypes', see Arumugam, Manimozhiyan et al., 'Enterotypes of the Human Gut Microbiome', *Nature* vol. 473, no. 7346 (12 May 2011): 174–80.

For the link between R gnavus and Crohn's disease, see Henke, Matthew T. et al., '*Ruminococcus Gnavus*, a Member of the Human Gut Microbiome Associated with Crohn's Disease, Produces an Inflammatory Polysaccharide', *Proceedings of the National Academy of Sciences of the United States of America* vol. 116, no. 26 (2019): 12672–77.

The Broad Institute at MIT has found the R gnavus phage and registered it as a patent. See https://www.broadinstitute.org/files/patents/WO2018195448A1.pdf (13 July 2021).

For faecal transplant as a cure, see Halkjær, Sofie Ingdam et al., 'Faecal Microbiota Transplantation Alters Gut Microbiota in Patients with Irritable Bowel Syndrome: Results from a Randomised, Double-blind Placebo-controlled Study', *Gut* vol. 67, no. 12 (2018): 2107–15.

For the discovery of crAssphage in 2014 in the human gut, see Dutilh, Bas E. et al., 'A Highly Abundant Bacteriophage Discovered in the Unknown Sequences of Human Faecal Metagenomes', *Nature Communications* vol. 5, no. 4498 (24 July 2014).

Scientists believe that crAssphage belongs to the largest bacteriophage family (Caudovirales) and lives in a wide range of environments, like termite guts, groundwater, salt pans, marine sediment, plant root systems and of course, our guts and faeces. See Camarillo-Guerrero, Luis F. et al., 'Massive Expansion of Human Gut Bacteriophage Diversity', *Cell* vol. 184, no. 4 (2021).

On the virome and microbiome of the gut of growing infants, children and adults, see Beller, Leen and Jelle Matthijnssens, 'What Is (Not) Known about the Dynamics of the Human Gut Virome in Health and Disease', *Current Opinion in Virology* vol. 37 (2019): 52–57.

For recent evidence on intestinal microbiota and its impact on allergies like asthma in adults, see Hufnagl, Karin et al., 'Dysbiosis of the Gut and Lung Microbiome Has a Role in Asthma', *Seminars in Immunopathology* vol. 42, no. 1 (2020): 75–93.

For specific co-evolution of immunity with viruses, see Koonin, Eugene V. and Yuri I. Wolf, 'Evolution of the CRISPR-Cas Adaptive Immunity Systems in

Prokaryotes: Models and Observations on Virus-host Co-evolution', *Molecular BioSystems* vol. 11, no. 1 (2015): 20–27.

On bacteriophage-adherence-to-mucus (BAM), see Barr, Jeremy J. et al., 'Subdiffusive Motion of Bacteriophage in Mucosal Surfaces Increases the Frequency of Bacterial Encounters', *Proceedings of the National Academy of Sciences of the United States of America* vol. 112, no. 44 (2015): 13675–80; and Almeida, Gabriel M.F. et al., 'Bacteriophage Adherence to Mucus Mediates Preventive Protection against Pathogenic Bacteria', *mBio* vol. 10, no. 6 (19 November 2010).

On the discovery of CRISPR and its role, see Barrangou, Rodolphe and Philippe Horvath, 'A Decade of Discovery: CRISPR Functions and Applications', *Nature Microbiology* vol. 2, no. 17092 (5 June 2017).

Nearly 40 per cent of bacteria have some sort of CRISPR system in their genomic sequences. See Karginov, Fedor V. and Gregory J. Hannon, 'The CRISPR System: Small RNA-guided Defense in Bacteria and Archaea', *Molecular Cell* vol. 37, no. 1 (2010): 7–19.

For the revised estimate of the extent that CRISPR-Cas systems confer microorganisms with immunity against viruses, see Burstein, David et al., 'Major Bacterial Lineages Are Essentially Devoid of CRISPR-Cas Viral Defence Systems', *Nature Communications* vol. 7, no. 10613 (3 February 2016).

For the landmark 2012 paper published in the journal *Science*, on the discovery by the 2020 Nobel Laureates for Chemistry, Professors Emmanuelle Charpentier and Jennifer Doudna on the action of CRISPR, see Jinek, Martin et al., 'A Programmable Dual-RNA-guided DNA Endonuclease in Adaptive Bacterial Immunity', *Science* vol. 337, no. 6096 (2012): 816–21.

For five means by which bacteria can protect themselves from viruses, see Ofir, Gal et al., 'DISARM Is a Widespread Bacterial Defence System with Broad Anti-phage Activities,' *Nature Microbiology* vol. 3, no. 1 (2018): 90–98.

On the microbiome-gut-brain axis is a two-way communication between bacteria and gut cells, see Dinan, Timothy G. and John F. Cryan, 'The Microbiome-Gut-Brain Axis in Health and Disease', *Gastroenterology Clinics of North America* vol. 46, no. 1 (2017): 77–89.

For an unbiased review of the efficacy of off-the-shelf probiotics, see Agans, Richard T. et al., 'Evaluation of Probiotics for Warfighter Health and Performance', *Frontiers in Nutrition* vol. 7, no. 70 (9 June 2020).

One review summarised that 'it may be plausible that consuming more fermented foods, improves the health of the gut microbiome and may stimulate the vagal afferents and functions of the brain'. See Bravo, Javier A. et al., 'Ingestion of Lactobacillus Strain Regulates Emotional Behavior and Central GABA Receptor Expression in a Mouse via the Vagus Nerve', *Proceedings of the National Academy of Sciences of the United States of America* vol. 108, no. 38 (2011): 16050–55.

9. A Virus Vanishes

The frost of 1683–84 was unparalleled in London's history. From early December 1683 until February 1684 the ice was as thick as it had ever been. Booths were set up on the frozen river where games and various entertainments took place. Coaches, sledges and sedan chairs are shown plying on the ice while a game of ninepins is in progress in the central foreground. Citizens stroll on the ice, carefully avoiding the gaping hole in the right foreground. Among the buildings in the background are, from left to right, St Clement Danes, Essex Buildings, Middle Temple Hall, Temple Stairs and King's Bench Walk. The diary entry of 1 January 1684 by John Evelyn, the chronicler of Restoration England read, 'The weather continuing intolerably severe, streets of booths were set upon the Thames, the air was so very cold and thick, as of many years there had not been the like. The smallpox was very mortal.' See *The Diary of John Evelyn*, in the 1879 edition by William Bray, available at https://archive.org/details/diaryofjohnevely01eveliala (19 July 2021).

This period witnessed the most intense volcanic eruptions since our ancestors left Africa, with several powerful and simultaneous eruptions. See Global Volcanism Program, 2013, Volcanoes of the World, v. 4.10.0 (14 May 2021), Venzke, E. (ed.), Smithsonian Institution, available at https://doi.org/10.5479/si.GVP.VOTW4-2013.

Cold conditions caused repeated collapse of agriculture and triggered cycles of war and disease, and each fed the other. An excellent overview is by Zhang, David D. et al., 'Global Climate Change, War, and Population Decline in Recent Human History', *Proceedings of the National Academy of Sciences of the United States of America* vol. 104, no. 49 (2007): 19214–19.

In a period of just seventy years or so (1440–1510), wine-tippling British took to quaffing whisky and gin, and guzzling warm beer. Read Fagan, Brian, *The Great Warming: Climate Change and the Rise and Fall of Civilizations*, Bloomsbury Publishing, 2020.

Plough horses were weaned away from coarse cereals, for which cultivation of forage crops like clover, beets and turnips was undertaken. See Tello, Enric et al., 'The Onset of the English Agricultural Revolution: Climate Factors and Soil Nutrients', *Journal of Interdisciplinary History* vol. 47, no. 4 (Spring 2017): 445–74.

Avoiding water became such a mania that by the early 1600s, white linen was considered a substitute to bathing. A wonderful history of hygiene is by Ashenburg, Katherine, *Clean: An Unsanitized History of Washing*, Profile Books, 2009. For two other interesting histories of sanitation and ablution in Europe and Britain, see DiPiazza, Francesca Davis, *Remaking the John. The Invention and Reinvention of the Toilet*, 21st Century Books, November 2014; and Horan, Julie L., *Porcelain God–A Social History of the Toilet*, Citadel New York, 2000.

The onset of the sweating disease started with the tell-tale sign of profuse sweating, followed by fever and lethargy, and impacted the brain and functioning of the nervous system. See Cheshire, William P. et al., 'Sudor Anglicus: An Epidemic

Targeting the Autonomic Nervous System', *Clinical Autonomic Research: Official Journal of the Clinical Autonomic Research Society* vol. 30, no. 4 (2020): 317–23.

In 1881, in his *Handbuch der Historisch-Geographischen Pathologie*, the German physician, August Hirsch, tabulated 194 outbreaks of sweating sickness in France and Spain from 1718 to 1874. See Hirsch, August, *Handbook of Geographical and Historical Pathology*, Volume 1: *Acute Infective Diseases*, translated from the second German edition by Ch. Creighton, New Sydenham Society, London, 1883.

Disease experts have argued on vetting possible pathogens, which include benign stomach viruses, while Catharine Arnold has labelled the sweating diseases as influenza. See Arnold, Catharine, *Pandemic 1918: The Story of the Deadliest Influenza in History*, Michael O'Mara Books, 2018.

What is remarkable is that some viral fevers despite persisting for a century or so vanish without a trace, and emerge in unexpected places, see Heyman, Paul et al., 'The English Sweating Sickness: Out of Sight, Out of Mind?', *Acta Medica Academica* vol. 47, no. 1 (2018): 102–16.

Raphael Holinshed's Chronicles, published in 1557, called the sweating sickness 'so sharp and deadly that the lyke was never hearde of to any manne's remembrance before that tyme'. Bear in mind this was written by people who had survived the Black Death. Available at the British Library https://www.bl.uk/collection-items/holinsheds-chronicles-1577 (13 July 2021).

On the history of the sudor in Austria and Germany, see Flamm, Heinz, 'Anno 1529 – der "Englische Schweiß" in Wien, die Türken um Wien', [Anno 1529 – the 'English Sweating Sickness' in Vienna, the Turkish Troops around Vienna], *Wiener medizinische Wochenschrift (1946)* vol. 170, nos 3–4 (2020): 59–70.

A comprehensive survey of major works on Danse Macabre was done by Sears, George Edward, *A Collection of Works Illustrative of Dance of Death*, New York, privately printed, 1889, available at https://collections.nlm.nih.gov/pdf/nlm:nlmuid-59031060R-bk (13 July 2021).

Nearly seventy books and papers have surveyed the possible cause of death of Mozart. I depended on Stafford, William, *Mozart's Death: A Corrective Survey of the Legends*, Palgrave Macmillan, UK, 1991, which provides a basis for the conditions of his death. More recent papers refer to military fever, starting with Cormican, Brendan, *Mozart's Death/Mozart's Requiem: An Investigation*, Belfast, N. Ireland: Amadeus Press, 1991. According to his death certificate housed in St Stephan's Church in Vienna, Wolfgang Amadeus Mozart died of hitziges frieselfieber (German for severe military fever or sweating sickness) in 1791.

Recent outbreaks of hantaviruses (initially labelled as a fever of unknown origin) affected the Navajo reservation community in New Mexico, USA in 1993. Medical researchers at the Queen Astrid Military Hospital in Brussels also reported a hantavirus-like fever in a journal article in 2013.

The story of sweating sickness epidemics is fascinating to understand how it influenced England's society and culture, before the virus disappeared. Henry Tidy, a British epidemiologist in his article in the *British Medical Journal* in 1945 concluded his observations on the sweating sickness as follows: 'Such is the disease which at one time was feared even above plague. We may bear in mind [Dr. Michael] Foster's warning: "We should be unwise to regard it as necessarily a disease nearing extinction." I am.' See Tidy, Henry, 'Sweating Sickness and Picardy Sweat, Correspondence', *BMJ*, (14 July 1945): 65–66.

The last case of the disease occurred in a single soldier in Picardy in 1918, and just like that, sweating fever vanished from Europe. See Foster, Michael, *Volume Dedicated to Osler on His Seventieth Birthday, Contributions to Medical and Biological Research* vol. 1, no. 52, Hoeber, New York, 1919.

There are forty-seven hantaviruses that have been identified thus far. It is believed that the ancestor of the hantaviruses diverged from a bunyavirus, and was passed down perhaps from an insect to an early mammal about 90 to 100 million years ago. This is reviewed in Plyusnin, Alexander and Tarja Sironen, 'Evolution of Hantaviruses: Co-speciation with Reservoir Hosts for more than 100 MYR', *Virus Research* vol. 187 (2014): 22–26.

'New' hantaviruses-related fevers are emerging in South East and South Asian regions. See Henttonen, Heikki et al., 'Recent Discoveries of New Hantaviruses Widen Their Range and Question Their Origins', *Annals of the New York Academy of Sciences* vol. 1149 (2008): 84–89.

Mice and voles are closer to humans, and a hantavirus outbreak may be in the offing warns Saxenhofer, Moritz et al., 'Secondary Contact between Diverged Host Lineages Entails Ecological Speciation in a European Hantavirus', *PLoS Biology* vol. 17, no. 2 (20 February 2019).

Outbreaks of hantaviruses, first thought to be a fever of unknown origin, affected the Navajo reservation community in New Mexico, USA in 1993 (labelled 1993 Four Corners hantavirus outbreak), see Chapman, L.E. and R.F. Khabbaz, 'Etiology and Epidemiology of the Four Corners Hantavirus Outbreak', *Infectious Agents and Disease* vol. 3, no. 5 (1994): 234–44. Medical researchers at the Queen Astrid Military Hospital in Brussels also reported a hantavirus-like fever in 2013. See Watson, Dionysios Christos et al., 'Epidemiology of Hantavirus Infections in Humans: A Comprehensive, Global Overview', *Critical Reviews in Microbiology* vol. 40, no. 3 (2014): 262–72.

10. Beauty

The notes to each chapter provide detailed references for the sources cited. Here, I provide suggested topical reading from books that I have found especially worthwhile. Goldgar, A., *Tulipmania: Money, Honor and Knowledge in the Dutch Golden Age*, University of Chicago Press, 2007, is a meticulous account and cultural history of the 1630 tulipmania in the Netherlands. Also *The Tulip: The Story of a Flower that Has Made Men Mad* by Pavord, A.; and MacKay, Charles, *Extraordinary Popular Delusions and the Madness of Crowds*, Harriman House, 2003; and Das, Mike, *Tulipomania: The Story of the World's Most Coveted Flower and the Extraordinary Passions It Aroused*, Crown, 2011. On fondness for tulips in Persia and in the Ottoman Empire, see Salzmann, Ariel, *The Age of Tulips: Confluence and Conflict in Early Modern Consumer Culture (1550-1730)*, in Donald Quataert (ed.), *Consumption Studies and the History of the Ottoman Empire 1550-1922: An Introduction*, State University of New York Press, December 1999. On design of gardens and the positioning of tulips in them, see Titley, Norah M., *Plants and Gardens in Persian, Mughal and Turkish Art*, British Library, 1979.

Tulips are native to the western slopes of the Tien Shan mountains in Kazakhstan and reaching central Crimea. There are about seventy-six species of tulips in the wild. See Maarten, J.M. Christenhusz et al., 'Tiptoe through the Tulips: Cultural History, Molecular Phylogenetics and Classification of Tulipa (Liliaceae)', *Botanical Journal of the Linnaean Society of London* (18 June 2013).

The ancestors of modern lilies emerged around 85 million years ago during the flower revolution during the heyday of the dinosaurs, in the period called Cretaceous. The ancestors of modern tulips were among the last to arrive on the scene about 21 million years ago somewhere in East Asia. See Kim, Jung Sung and Joo-Hwan Kim, 'Updated Molecular Phylogenetic Analysis, Dating and Biogeographical History of the Lily Family (Liliaceae: Liliales)', *Botanical Journal of the Linnaean Society* vol. 187 (2018): 579–93.

Anna Pavord remarks that 'it was inevitable that tulipomania would be followed by an equally intense hatred of the flower', and states that vanitas still lifes (which actually dated from the later sixteenth century) were a result of the 'chastening experience' of the tulip crash. See Pavord, Anna, *The Tulip*, Bloomsbury Press, 2000. Paul Taylor is perhaps less guilty of this cliché than some, but he still remarks that Jacob Gerritsz Cuyp's unusual picture of a bed of tulips from 1638 'must have been so many hot needles in the flesh of someone bankrupted in the previous year'. See Taylor, Paul, *Dutch Flower Painting 1600–1720*, Yale University Press, 1995. For a more comprehensive survey of Dutch and Flemish flower painting, see Segal, Sam and Klara Alen, *Dutch and Flemish Flower Pieces: Paintings, Drawings and Prints up to the Nineteenth Century*, Brill Academic Publishing, October 2020.

Simon Schama argues in his classic survey *The Embarrassment of Riches*, that the seventeenth-century Dutch reviled their wealth even as they piled it up, and tulips became its leitmotif.

Another factor that could have impacted the bubble was the emergence of plague in the winter of 1637 in Haarlem. Outbreaks of bubonic plague emerged in the city from the summer of 1635 until the end of 1636, and there were panic sales and frenetic buying in Haarlem, Leiden and Amsterdam. See Curtis, Daniel, 'Was

Plague an Exclusively Urban Phenomenon? Plague Mortality in the Seventeenth-Century Low Countries', *Journal of Interdisciplinary History* vol. 47, no. 2 (August 2016): 139–70.

In 1931, the unknown viral agents that caused diseases in potatoes were labelled potato virus X, and viral agents in potatoes and tulips were called potato virus Y (hence *potyvirus*). It was transmitted by infected sap—from potatoes to peaches and apples to tulips, and not necessarily following a definite pathway. Since its discovery, the 214 potyviruses is therefore a large plant virus genus and is one of the most important economically because of the yield and quality losses it causes in a wide range of crops worldwide. See Gibbs, Adrian J. et al., 'The Potyviruses: An Evolutionary Synthesis Is Emerging', *Viruses* vol. 12, no. 2 (22 January 2020): 132.

The early experiments by Dorothy Cayley, a mycologist at John Innes Centre, UK, suggested the degree of breaking was proportional to the amount of infected tissue introduced and was published in the *Annals of Applied Biology* in November 1928. See Cayley, D.M., '"Breaking" in Tulips', *Annals of Applied Biology* vol. 15, no. 4 (November 1928): 529–39. A follow-up paper in 1932 confirmed her findings, and since then Cayley has been credited as the scientist who solved the puzzle of 'breaking' in tulips. See Cayley, D.M., '"Breaking" in Tulips II', *Annals of Applied Biology* vol. 19, no. 2 (May 1932): 153–72. Sadly, Cayley rarely finds a mention among the pioneers of plant pathology or virology. Her contribution to mycology, the study of fungi, was immense. She described twenty-eight new species of fungi during her time, and mentored a cohort of mycologists and plant pathologists during her tenure at the centre.

The potyvirus interferes with the production of pigments in flowers and fruits. See Gibbs, Adrian J. et al., 'The Prehistory of Potyviruses: Their Initial Radiation Was during the Dawn of Agriculture', *PLoS one* vol. 3, no. 6 (25 June 2008). By mixing genes that code its coat protein and a helper gene that makes these viruses compatible for transmission through migrating aphids, several varieties of potyviruses have evolved into distinct virus species. See Gibbs, Adrian J. et al., op. cit.

By using potyviruses that infect beet, it was found that sugar beet emerged as a crop over the past three centuries while leaf beet was used as a vegetable for at least 2000 years. Read Mohammadi, Musa et al., 'An Iranian Genomic Sequence of Beet Mosaic Virus Provides Insights into Diversity and Evolution of the World Population', *Virus Genes* vol. 54, no. 2 (2018): 272–79.

Virologists have found that a related virus that infects turnips (turnip mosaic virus [TuMV]) probably originated from a virus of wild orchids in Germany and, while adapting to wild and domestic brassicas, spread via Southern Europe to Asia Minor no more than 700 years ago, and later spread into Iran and eastwards. The timescale of the spread of TuMV confirms the period when agriculture took root in these countries. For this, see Yasaka, Ryosuke et al., 'The Timescale of Emergence and Spread of Turnip Mosaic Potyvirus', *Scientific Reports* vol. 7, no. 1 (26 June 2017): 4240.

Other viruses also induce different forms of breaking of the flower colour apart from the leaf symptoms like the tobacco rattle tobravirus, tobacco necrosis necrovirus and lily symptomless carlavirus. A combination of factors like acidity, environmental stress to the plant and bulb, and co-infection of two or more viruses, cause the shades, shape and streaks in the petals to become very distinct. The resplendent ivy-leaf geranium develops pink-and-white streaks when it gets infected by a virus of another flower, the Pelargonium flower break virus. See Valverde, Rodrigo A. et al., 'Viruses that Enhance the Aesthetics of Some Ornamental Plants: Beauty or Beast?', *Plant Disease* vol. 96, no. 5 (2012): 600–11.

In Australia a mild strain of GLR increased berry weight and volume; in Japan the acid content reduced in juices from virus-infected Merlot grapevines; virus infection in Cabernet Sauvignon grapes lowered acid yield and sharpness; in Italy GLRV-infected Dolcetto vines produced wines that have soft bouquet but higher plum aroma, astringency, body and violet colour. See Reynolds, Andrew, 'The Grapevine, Viticulture, and Winemaking: A Brief Introduction', Chapter in B. Meng et al. (eds), *Grapevine Viruses: Molecular Biology, Diagnostics and Management*, Springer AG, July 2017.

Farmers in Japan are able to use ALSV vectors to shorten the germination time of apples and pears and speed up the breeding of grape for desired traits. Maeda, Kiyoaki et al., 'Virus-Induced Flowering by Apple Latent Spherical Virus Vector: Effective Use to Accelerate Breeding of Grapevine', *Viruses* vol. 12, no. 1 (7 January 2020): 70.

11. How a Virus Saved a Giant

The most comprehensive survey of the chestnut blight is presented by Freinkel, Susan, *American Chestnut: The Life, Death, and Rebirth of a Perfect Tree*, University of California Press, 2009.

The fungus was first described as *Diaporthe parasitica* Murrill and later transferred to the genus *Endothia*. In 1978, it received its current name, *Cryphonectria parasitica* (Murr.) Barr, 1978. *Cryphonectria parasitica* has become a textbook example of an introduced pathogen that has caused devastating disease epidemics in native tree species.

Frost, Robert, *The Complete Poems of Robert Frost*, New York: Holt, Rinehart and Winston, 1964.

Other major forest trees diseases include white pine blister rust, and oak blight but none has caused decimation of forests like the eastern American chestnut forests. By 1920, all hope was lost and efforts to save the chestnut were abandoned as is mentioned in MacDonald, W.L. and D.W. Fulbright, 'Biological Control of Chestnut Blight: Use and Limitation of Transmissible Hypovirulence', *Plant Disease* vol. 75 (1991) 656–61.

By 1917, almost all mature chestnut trees in Connecticut had died, and by 1924, all hope was lost and efforts to save the chestnut were abandoned. By the 1930s, forests appeared like an apparition of their former glory. By 1940, of the estimated 4 billion trees, over 3.5 billion mature trees were lost to the blight. The chestnut blight singularly was the greatest ecological crisis in North America since the great extinction of megafauna mammals, the mammoths, mastodons, giant beaver among others, that occurred between 12,000 and 9000 years ago, and was caused due to overkill by human settlers and changing climate. See Broughton, Jack M. and Elic M. Weitzel, 'Population Reconstructions for Humans and Megafauna Suggest Mixed Causes for North American Pleistocene Extinctions', *Nature Communications* vol. 9, no. 1 (21 December 2018): 5441.

With the dominant trees dead and gone, many birds that depended on them for nesting like Cooper's hawk, woodpeckers and cerulean warbler also declined. Bear and racoon populations thinned down. See Hill, James M., 'Wildlife Value of *Castanea dentata*: Past and Present, the Historical Decline of the Chestnut and Its Future Use in Restoration of Natural Areas', *Proceedings of the International Chestnut Conference* vol. 1 (1994): 186–94.

Native American folklore and several cultural practices linked to the chestnut disappeared. Davis, D.E., 'Historical Significance of American Chestnut to Appalachian Culture and Ecology', *Proceedings of the Conference on Restoration of American Chestnut to Forest Lands*, Steiner, K.C. and J.E. Carlson (eds), 2005.

At least a fourth of all trees in the forest were chestnuts and their girth occupied as a proportion, more than a third of all trees in the forests. Elliott, K.J. and W.T. Swank, 'Long-term Changes in Forest Composition and Diversity Following Early Logging (1919–1923) and the Decline of American Chestnut (*Castanea dentata*)', *Plant Ecology* vol. 197, no. 2 (2008): 155–72.

Together these altered the lives of creatures of the chestnut-dominated woodlands, and people who have lived around these forests. Before their demise, the American chestnut dominated the hardwood forests of the eastern US.

In 1950, Bilbao in Spain reported an outbreak of canker in chestnut groves, spread purportedly through an infection carried by Japanese chestnut seeds. In 1951, Spanish foresters decided to axe the old growth stand of chestnut trees to prevent its spread. See this paper which describes the repeated appearance of the fungus in Spain. Bascón, J. et al., 'First Report of Chestnut Blight Caused by Cryphonectria Parasitica in a Chestnut Orchard in Andalusia (southern Spain)', *Plant Disease* vol. 98, no. 2 (2014): 283–84.

A good recent overview is presented by Brian Lovett which is available at https://asm.org/Articles/2020/May/Restoring-the-American-Chestnut-with-a-Virus-and-B (13 July 2021).

In 2013, one paper labelled American chestnuts of the eastern hardwood forests as 'functionally extinct'. See Säterberg, Torbjörn et al., 'High Frequency of Functional Extinctions in Ecological Networks', *Nature* vol. 499, no. 7459 (2013): 468–70.

Between 1955 and 1974 chestnut production declined from 2,37,000 tonnes to just 63,500 tonnes (in 1935 before the onset of the blight it was 3,40,000 tonnes). See Mittempergher, L., 'The Present Status of Chestnut Blight in Italy', in *Proc. Am. Chestnut Symp*, MacDonald, W.L., F.C. Cech, J. Luchok and C. Smith (eds), Morgantown: West Virginia University Books, 1982, pp. 34–37.

Early papers which reported on hypovirulence include Van Alfen, N.K. et al., 'Chestnut Blight: Biological Control by Transmissible Hypovirulence in Endothia Parasitica', *Science* vol. 189, no. 4206 (1975): 890–91; and Hansen, D.R. et al., 'Naked dsRNA Associated with Hypovirulence of *Endothia Parasitica* Is Packaged in Fungal Vesicles', *Journal of General Virology* vol. 66, no. 12 (1985).

The role of a virus in hypovirulence was established in 1980. See Dodds, J.A., 'Association of Type 1 Viral-like dsRNA with Club-shaped Particles in Hypovirulent Strains of Endothia Parasitica', *Virology* vol. 107, no. 1 (1980): 1–12.

Biraghi continued to collect evidence of death and early signs of recovery from other countries and was persistent in his advocacy for concerted action across Europe. See US Dept of Agriculture, *Plant Disease Reporter* vol. 36, no. 3 (15 March 1952).

The contribution of Biraghi in plant pathology can be found in his obituary. See Anon, 'A. Biraghi, In memoriam', *Caryologia* vol. 24, no. 4 (1971).

It was Grente and Berthelay-Sauret who presented the idea of biological control using the virus. See Grente, J. and S. Berthelay-Sauret, 'Biological Control of Chestnut Blight in France', in *Proceedings of the American Chestnut Symposium* (1978): 30–34, MacDonald, W.L., F.C. Cech, J. Luchok & C. Smith (eds), Morgantown: West Virginia University Press; and Grente, J., 'Les Variants Hypovirulents de l'*Endothia parasitica* et la Lutte Biologique Contre le Chancre du Châtaignier', Université de Bretagne Occidentale, Brest, France, Dissertation (1981). Sustained field experiments by Grente and Berthelay-Sauret showed that the trees recovered not only because HV strain arrested the spread of the V strain, but it was also able to convert the V to an HV strain by transmission of a 'cytoplasmic determinant'.

In Wisconsin, a twenty-year-old trial of inoculating HV strains in established cankers has nearly doubled survival, and most trees have regained their canopy space. See Double, M.L. et al., 'Evaluation of Two Decades of *Cryphonectria parasitica Hypovirus* Introduction in an American Chestnut Stand in Wisconsin', *Phytopathology* vol. 108, no. 6 (2018): 702–10.

For the field trials of the American Chestnut Foundation (ACF) of crossing the Chinese chestnut with the American one and creating hybrids, and the

Canadian Chestnut Council, see their respective websites. There are also small successes thanks to the tireless efforts of universities and chestnut societies. Success is afoot.

Asian invader, the Oriental chestnut gall wasp (*Dryocosmus kuriphilus*) has entered Europe and North America and can potentially create havoc. The wasp lays its eggs in the fleshy shoots and leaves of the tree and its larvae feast on it. The graver concern is that the wasp is a prolific carrier of the blight fungus. See Pérez-Sierra, A. et al., 'First Report of *Cryphonectria parasitica* on Abandoned Galls of *Dryocosmus kuriphilus* on Sweet Chestnut in the United Kingdom', *New Disease Reports* vol. 41, no. 34 (2020).

Mycologists and virologists are teaming up to prospect for viruses that can stymie dreaded fungal epidemics caused by Dutch elm disease, white pine blister rust, and oak blight. See Sharma, Mohit et al., 'Mycovirus Associated Hypovirulence, a Potential Method for Biological Control of *Fusarium* Species', *Virusdisease* vol. 29, no. 2 (2018): 134–40.

12 Zombies

Expert: 'No Test Proves that Coffee and Onions Can Cure [Ebola]', *Xinhuanet*, 9 August 2014, available at https://web.archive.org/web/20201117135321/http://news.sina.com.cn/w/2014-08-09/122530658162.shtml (14 July 2021).

The baike webpage (baike.com/wiki/丧尸病毒?view_id=3ui38g1vgrq000) was removed but a screengrab of the page is available here: https://archive.org/details/baike-ebola-synonymous-with-zombie-virus (14 July 2021). To understand the frenzy building up on threats from Ebola before the Beijing Olympics, see Liu, Kui et al., 'Chinese Public Attention to the Outbreak of Ebola in West Africa: Evidence from the Online Big Data Platform', *International Journal of Environmental Research and Public Health* vol. 13, no. 8 (4 August 2016): 780.

On the *New Dawn*, Liberia's report on the emergence of zombies, available at https://www.ibtimes.co.in/ebola-zombies-liberian-newspaper-claims-victims-are-rising-dead-610439 (14 July 2021).

'Any plans to rebuild and return England to its pre-attack glory would be led by the Cabinet Office, and thus any pre-planning activity would also take place there.' Quoted in the *Telegraph*, available at https://www.telegraph.co.uk/news/newstopics/howaboutthat/9721072/Britain-is-well-prepared-to-fight-apocalyptic-zombie-invasion.html (14 July 2021).

CONPLAN 8888 lists the use of hand sanitisers, UV rays, and how the army will need to target the different kinds of zombies like flesh-eaters with 'firepower to the head'. Read this interesting document available at https://www.stratcom.mil/Portals/8/Documents/FOIA/CONPLAN_8888-11.pdf?ver=2016-10-17-114016-887 (14 July 2021).

US Centers for Disease Control and Preparedness (CDC) since October 2016 has a dedicated page on how you should survive a zombie threat, available at https://www.cdc.gov/cpr/zombie/index.htm. (14 July 2021).

The Right To Information (RTI) question posed to the Ministry of Home Affairs on a possible zombie attack, available at https://www.thequint.com/neon/social-buzz/man-files-rti-is-government-ready-for-zombie-attack-gets-response-home-affairs-will-smith (14 July 2021).

T. gondii have been associated with greater probability of meeting with car accidents, suffering from mental illness, neuroticism, drug abuse and suicide. On a more positive note, a study published in *Proceedings of the Royal Society* found those infected with *T. gondii* possess stronger entrepreneurial drive. See Johnson, Stefanie K. et al., 'Risky Business: Linking *Toxoplasma gondii* Infection and Entrepreneurship Behaviours across Individuals and Countries', *Proceedings of the Royal Society, Series B: Biological Sciences* vol. 285, no. 1883 (25 July 2018).

The Zombie Research Society, available at https://zombieresearchsociety.com/archives/36443, has eminent scientists, film-makers, writers and public health experts who imagine scenarios of a possible rise of the zombies (14 July 2021).

This co-evolution of PDV has occurred over the past 100 million years with the 18,000 or so parasitic wasps that live today; each wasp has its own specific virus. Somewhere during this journey of mutualism and gene swapping the identities of the virus and the wasp have become blurred. The integration is so complete that it is difficult to differentiate between virus and host, and genes get passed on to the next generations. See Herniou, Elisabeth A. et al., 'When Parasitic Wasps Hijacked Viruses: Genomic and Functional Evolution of Polydnaviruses', *Philosophical Transactions of the Royal Society of London, Series B, Biological Sciences* vol. 368, no. 1626 (12 August 2013).

On the takeover by fungi on insects with ants as a model, see Hughes, David P. et al., 'Behavioral Mechanisms and Morphological Symptoms of Zombie Ants Dying from Fungal Infection', *BMC Ecology* vol. 11, no. 13 (9 May 2011).

Roughly 15 to 20 per cent of all insect species are parasitoids, insects that reproduce by laying their eggs in a developing stage of another insect, the host. Additionally, parasitoids play a vital role in controlling the populations of insects that are crop pests or have other negative impacts. The population of viruses, wasps and cockroaches are dependent on myriad factors and each other. Two textbooks that discuss this elegantly are Hassell, M.P., *The Dynamics of Arthropod Predator-Prey Systems*, Princeton University Press, 1978; and Anderson, R.M. and R.M. May, *Infectious Diseases of Humans: Dynamics and Control*, Oxford University Press, 1991.

Charles Darwin in an 1860 letter to his friend, the botanist Asa Gray, said, 'I cannot persuade myself that a beneficent & omnipotent God would have designedly created the *Ichneumonidæ* [the parasitoid wasp] with the express intention of their feeding within the living bodies of caterpillars, or that a cat should play with mice.' Available at https://www.darwinproject.ac.uk/letter/DCP-LETT-2814.xml (14 July 2021).

Legendary Swiss artist H.R. Giger, director Ridley Scott, and writers Dan O'Bannon and Ronald Shusett made wasps their inspiration to create one of the most enduring extra-terrestrial monsters in film for the classic sci-fi horror, *Alien* (1979). Their alien is a terrifying insect-like creature which is protected by a silicon-based external skeleton and contains acidic blood that can cut through human flesh and the metal of the spaceship. The film depicts a paralysing terror of an alien mother in search of a human host, and once she enters a spaceship, she stings and paralyses the ship's crew, infecting them with an egg that hatches within and eats its hosts, before it bursts out in most gruesome ways—all of which are reminiscent of a wasp's life cycle. Available at https://alienseries.wordpress.com/2012/10/18/the-insect-influence/ (14 July 2021).

Pleasant floral aroma that is commonly found in essential oils—all of these are released in tiny quantities by tree canopies when they want to attract insects. On how insects get attracted by trees which emit volatile compounds, see Lämke, Jörn S. and Sybille B. Unsicker, 'Phytochemical Variation in Treetops: Causes and Consequences for Tree-insect Herbivore Interactions', *Oecologia* vol. 187, no. 2 (2018): 377–88. Three chemicals in particular—benzyl nitrile, a mild cyanide poison; acetic acid, commonly known as vinegar, and 2-phenylethanol—together create these floral aromas.

The virus manipulates the caterpillar's physiology by stopping it from moulting, and hijacks its brain to seek light, a phenomenon not seen in uninfected ones, but the heavy and turgid virus-laden caterpillars head laboriously for the treetop. For this, see van Houte, Stineke et al., 'Baculovirus Infection Triggers a Positive Phototactic Response in Caterpillars to Induce "Tree-top" Disease', *Biology Letters* vol. 10, no. 12 (2014); and Rebolledo, Dulce et al., 'Baculovirus-Induced Climbing Behavior Favors Intraspecific Necrophagy and Efficient Disease Transmission in Spodoptera Exigua', *PLoS one* vol. 10, no. 9 (24 September 2015).

On how wasp offspring depend upon the virus to overcome the caterpillar's immune system and the mutualism and blurred identities of the virus and wasp, see Herniou, Elisabeth A. et al., op. cit.

On the grass-fungus-virus relationship, see Roossinck, Marilyn J., 'Move Over, Bacteria! Viruses Make Their Mark as Mutualistic Microbial Symbionts', *Journal of Virology* vol. 89, no. 13 (2015): 6532–35.

The poor plant, however, spends extra energy to produce the chemical signal to no avail and presumably also by becoming more obvious to potential enemies. See Mauck, Kerry E. et al., 'Deceptive Chemical Signals Induced by a Plant Virus Attract Insect Vectors to Inferior Hosts', *Proceedings of the National Academy of Sciences of the United States of America* vol. 107, no. 8 (2010): 3600–05.

By eating a sick individual, only a few caterpillars get infected, and most survive, and this reduces the spread of the baculovirus infection. See Van Allen, Benjamin G. et al., 'Cannibalism and Infectious Disease: Friends or Foes?', *American Naturalist* vol. 190, no. 3 (2017): 299–312.

The 'Cordyceps Brain Infection' fungus grows while the host is still alive, with 'the Infected' (as the game develops, players label their zombies) undergoing four stages of infection and in some way becoming zombified. The parasitic infection spreads through bites only to living hosts though infected dead hosts can release spores. Details of the game are available at https://en.wikipedia.org/wiki/The_Last_of_Us_Part_II and https://thelastofus.fandom.com/wiki/Cordyceps_Brain_Infection (14 July 2021).

The human placenta is extremely resistant to infection because these cells release certain ERV-derived proteins that stay with us in our gut and airway to protect us when we take our first breath and our first feed, see Chuong, E.B., 2018, op. cit.

The role of the virus within the insect is not fully understood and if there is any further manipulation other than making it aggressively seek new treetops and plants. See Safari, Maliheh et al., 'Manipulation of Aphid Behavior by a Persistent Plant Virus', *Journal of Virology* vol. 93, no. 9 (17 April 2019).

Viruses can also potentially increase their opportunity for transmission by providing rewards to the host plants that they infect. This may be conditional; for example, by aiding the plant's survival under weather shocks like drought or cold but surprisingly also by helping them flower gregariously in more favourable times, and advertising to attract pollinators. See Carr, John P. et al., 'Viral Manipulation of Plant Stress Responses and Host Interactions with Insects', *Advances in Virus Research* vol. 102 (2018): 177–97.

The first vampire frenzy miffed the otherwise sangfroid Voltaire so much that he wrote that 'nothing was spoken of but vampires, from 1730 to 1735'. See Voltaire, *The Works of Voltaire*, Vol. VII (Philosophical Dictionary Part 5), 1764, available at https://oll.libertyfund.org/titles/voltaire-the-works-of-voltaire-vol-vii-philosophical-dictionary-part-5 (14 July 2021).

A study published in the *Lancet Psychiatry* has found that almost one in five people who have had COVID-19 develop some form of mental illness, like anxiety disorders, insomnia, and dementia among others, within three months of testing positive. See Taquet, Maxime et al., 'Bidirectional Associations between COVID-19 and Psychiatric Disorder: Retrospective Cohort Studies of 62,354 COVID-19 Cases in the USA', *Lancet Psychiatry* vol. 8, no. 2 (2021): 130–40.

On the virus-insect host-tree partnership, read Safari, Maliheh, 2019, op. cit., and Mauck, K.E., 2010, op. cit.

The placenta and foetus are a tempting proposition for any parasite but why so few are able to cause any real damage, is because of our much-maligned hero—the virus. The surface of the placenta is lined with a fused-cell layer that is made by the endoretrovirus. In fact, it is the endoretrovirus, that in the first place enables us to have a placenta. Specific virus-derived proteins called syncytin have made the human placenta extremely resistant to infection. They stay with us in our gut and airway to protect us when we take our first breath and our first feed, Chuong, E.B., 2018, op. cit. Despite this protection, TORCH-z organisms can reach the foetus, and the mother and child need to be protected from these specific organisms. But in most cases, the inbuilt viruses-induced protection safeguards us from potential microbial invasions.

For a good review of film, culture, politics and zombies, see Wasik, Bill and Monica Murphy, *Rabid: A Cultural History of the World's Most Diabolical Virus*, Penguin Books, 2013.

An excellent online simulation named Zombietown USA gives you the controls to assess point of outbreak, available at http://mattbierbaum.github.io/zombies-usa/ (14 June 2021).

Researchers from Brazil have found that bones from an 80-million-year-old plant-eating dinosaur have several worms that bored through its bones and perhaps its skin. Due to the gruesome nature of the injuries, the scientists called this creature a 'Zombie Dino'. See Aureliano, Tito et al., 'Blood Parasites and Acute Osteomyelitis in a Non-avian Dinosaur (Sauropoda, Titanosauria) from the Upper Cretaceous Adamantina Formation, Bauru Basin, Southeast Brazil', *Cretaceous Research* vol. 118 (February 2021).

Human infectious diseases are caused by more than 1400 pathogen species with considerable diversity in lifestyles. Almost all pathogens of newly emerging diseases come from animal reservoirs. Most are viruses, especially RNA viruses. See Woolhouse, Mark and Rustom Antia, 'Emergence of New Infectious Diseases', in Stephen C. Stearns and Jacob C. Koella (eds) *Health and Disease*, November 2007. For a broad description of the 219 viruses that infect humans, see Woolhouse, Mark et al., 'Human Viruses: Discovery and Emergence', *Philosophical Transactions of the Royal Society of London, Series B, Biological Sciences* vol. 367, no. 1604 (2012): 2864–71.

Gómez-Alonso, J., 'Rabies: A Possible Explanation for the Vampire Legend', *Neurology* vol. 51, no. 3 (1998): 856–59.

See the 2006 Kiwi film, *Black Sheep*, available at https://www.imdb.com/title/tt0779982/ (14 July 2021).

For another interesting article on rabies as a source of vampires and other myths, see Maas, R.P.P.W.M. and P.J.G.M. Voets, 'The Vampire in Medical Perspective: Myth or Malady?', *QJM : Monthly Journal of the Association of Physicians* vol. 107, no. 11 (2014): 945–46.

For a short news item on the prospect of rabies increasing libido, see 'Humping like Rabids', *Playboy* (1 March 1999).

Letter from Darwin to Asa Gray, see Darwin Correspondence Project, 'Letter no. 2814', available at https://www.darwinproject.ac.uk/letter/DCP-LETT-2814.xml (14 July 2021).

On 'manipulation hypothesis', read Steffen, Hans-Michael, 'Wie beeinflussen Parasiten das Verhalten ihres Wirts? Die parasitäre Manipulationshypothese', [Behavioural Changes Caused by Parasites: The Parasite Manipulation Hypothesis], *Deutsche Medizinische Wochenschrift (1946)* vol. 145, no. 25 (2020): 1848–54.

On mortuary feasts by Fore people, see Whitfield, Jerome T. et al., 'Mortuary Rites of the South Fore and Kuru', *Philosophical Transactions of the Royal Society of London, Series B, Biological Sciences* vol. 363, no. 1510 (2008): 3721–24.

The Fore population faced a serious population bottleneck and stared at extinction. See Alpers, Michael P., 'Review. The Epidemiology of Kuru: Monitoring the Epidemic from Its Peak to Its End', *Philosophical Transactions of the Royal Society of London, Series B, Biological Sciences* vol. 363, no. 1510 (2008): 3707–13.

On the decline in cannibalism and prevalence of kuru in the mid-1950s, see Liberski, Pawel P. et al., 'Kuru: Genes, Cannibals and Neuropathology', *Journal of Neuropathology and Experimental Neurology* vol. 71, no. 2 (2012): 92–103.

Coining of prion, 1982, by Zabel, Mark D. and Crystal Reid, 'A Brief History of Prions', *Pathogens and Disease* vol. 73, no. 9 (2015).

On CJD and its 'evolution', see Collinge, J. et al., 'Molecular Analysis of Prion Strain Variation and the Aetiology of "New Variant" CJD', *Nature* vol. 383, no. 6602 (1996): 685–90.

University College London studies on CJD, available at https://www.ucl.ac.uk/prion/ (14 July 2021).

On minute alteration in one gene that makes it pathogenic, see Kim, Hyeon O. et al., 'Prion Disease Induced Alterations in Gene Expression in Spleen and Brain prior to Clinical Symptoms', *Advances and Applications in Bioinformatics and Chemistry: AABC* vol. 1 (2008): 29–50.

On genetically modifying mice to produce only the variant (valine) protein, see Wadsworth, J.D.F. et al., 'Review: Contribution of Transgenic Models to Understanding Human Prion Disease', *Neuropathology and Applied Neurobiology* vol. 36, no. 7 (2010): 576–97.

13 Enemy's Enemy

Maagh is the name of the tenth Hindu month and corresponds to the month of January–February. It occurs when the full moon is nearest to the constellation Magha α Leonis (Regulus). The cholera outbreak was at the Maagh Mela in

Allahabad, a fortnight-long festival of Hindus which attracted over 3 million pilgrims every day. See Anon, 'Allahabad Magh Mela of 1894', *Indian Medical Gazette* vol. 29, no. 3 (March 1894): 97–98.

On the seasonality and periodicity of the cholera epidemic along the Ganga and other rivers, see Arnold, David, 'The Ecology and Cosmology of Disease in the Banaras Region', in Sandria B. Freitag (ed.), *Culture and Power in Banaras: Community, Performance, and Environment,* 1800–1980, Berkeley: University of California Press, 1989. This can also be seen at UC Press E-Books Collection, available at http://ark.cdlib.org/ark:/13030/ft6p3007sk/ (14 July 2021). In particular, see the graphic on cholera deaths among pilgrims dispersing from Allahabad Kumbh Mela in February 1894.

On the cholera cycle, see Val, Marie-Eve et al., 'The Single-stranded Genome of Phage CTX Is the Form Used for Integration into the Genome of *Vibrio Cholerae*', *Molecular Cell* vol. 19, no. 4 (2005): 559–66.

A comprehensive review of Hankin's work and the mythic nature of the Ganges is in Kochhar, Rijul, 'The Virus in the Rivers: Histories and Antibiotic Afterlives of the Bacteriophage at the Sangam in Allahabad', *Royal Society Journal of the History of Science* (22 July 2020).

For a concerted plan to add this purple chemical to wells and instruct army cantonment officers and officers new to India to eat 'purple salad' to avoid diarrhoea, see Hankin, E.M., 'The Bactericidal Action of the Waters of the Jamuna and Ganges Rivers on Cholera Microbe', *Ann Inst Pasteur* vol. 10 (1896): 511–23.

In 1896 Hankin described the antibacterial activity of a then unknown source in the Ganges and Jamuna rivers in India. He noted, 'It is seen that the unboiled water of the Ganges kills the cholera germ in less than 3 hours. The same water, when boiled, does not have the same effect. On the other hand, well water is a good medium for this microbe, whether boiled or filtered.' He suggested that it was responsible for limiting the spread of cholera. See Hankin, E., 1896, op. cit.

Due to his inability to explain how infusing Ganga water cleared pathogens from infected wells, Hankin proposed that British officers use potassium permanganate as a general antiseptic against bacteria, and recommended that cantonments serve their new officers 'purple salad' to avoid diarrhoea. The findings of Hankin were critical considering what was to come in finding the agent or 'Materia X' and a 'cure to the cholera epidemic'. Hankin, 1896, op. cit.

Mark Twain, *More Tramps Abroad*, London: Chatto & Windus, 1897, pp. 343–44.

D'Hérelle continued to demonstrate the efficacy of the yet unseen but felt effects of phage therapy in France, and it was gradually tried by other scientists across Europe. The first patient was treated at the department of paediatrics at the Hôpital des Enfants-Malades in Paris on 2 August 1919. The patient was a twelve-year-old boy hospitalised with severe dysentery (ten to twelve bloody stools per day). After taking pre-treatment stool samples for microbiological analysis, d'Hérelle orally administered 2 ml of his most potent anti-dysentery phage preparation. The patient's condition rapidly improved after the ingestion of phage. Three additional patients with bacterial dysentery were successfully treated shortly afterwards: three brothers (aged three, seven and twelve years old) were admitted in grave condition in September 1919, after their sister had died at home due to the illness. In 1921, there was another report of successful use of phage in a hospital from Louvain in a French medical journal. Two doctors had isolated a staphylococcal phage and injected it in the local region of boils on the skin of a man suffering from swelling, pain and fever. A single injection cured the patient. Gradually, several hospitals in France and the rest of Europe took to phage therapy. Despite his major discovery, d'Hérelle was yet to make a place for himself at the Pasteur Institute.

On the cholera cycle, see Colwell, R.R. and A. Huq, 'Global Microbial Ecology: Biogeography and Diversity of Vibrios as a Model', *Journal of Applied Microbiology* vol. 85, Suppl. 1 (1998): 134S–37S.

D'Hérelle was aware of the action of the phage, although he did not know of the intricacies of the balance between lytic and lysogenic phages, and their role in the (cholera) disease cycle. For d'Hérelle's discovery of a microbe that was 'antagonistic' to bacteria and caused death in liquid culture and formed 'plaques', see d'Hérelle, F., 'Sur un Microbe Invisible Antagoniste des Bacilles Dysenteriques', *C.R. Acad. Sci. Ser. D* vol. 165 (1917): 373–75; and d'Hérelle, F., 'The Bacteriophage', *Science News*, vol. 14 (1949): 44–59.

For an overview of d'Hérelle's contributions, see Summers, W.C., *Felix d'Hérelle and the Origins of Molecular Biology*, New Haven, CT: Yale University Press, 1999.

For a review of the 1932 investigation by three scientists who compared Twort and d'Hérelle's material and their conclusions, see Flu, P.C. and E. Renaux, 'Le phénomène de Twort et la bactériophagie', *Annales de l'Institut Pasteur* vol. 48 (1932): 15–18.

The summer of 1919 monograph by d'Hérelle of phage use as a prophylaxis against *Bacillus gallinarum* infection of chickens or avian typhosis was blocked by Bordet for release and only when his tenure ended, was published later as d'Hérelle, F., *Le Bactériophage: Son Rôle dans l'Immunité*. Paris: Masson et Cie, 1921.

On the Bacteriophage Enquiry in India, under the patronage of the Indian Research Fund Association to study the feasibility of phage therapy for cholera and plague epidemics in India, see Asheshov, I.N., J. Asheshov, S. Khan and M.N. Lahiri, 'Bacteriophage Enquiry: Report on the Work during the Period from 1 January to 1 September, 1929', *Indian Journal of Medical Research* vol. 17, no. 971 (1930); d'Hérelle, F., R.H. Malone and M.N. Lahiri, 'Studies on Asiatic Cholera', *Indian Medical Research Memoirs* (1930) Memoir 14; and a review, Summers, 1993, op. cit.

From the initial reports from India in the 1920s and 1930s, Asheshov, Ignor N. et al., 'The Treatment of Cholera with Bacteriophage', *Indian Medical Gazette* vol. 66, no. 4 (1931): 179–84, it seems consistently observed that the severity and duration of cholera symptoms and the overall mortality from the disease were reduced in patients given cholera-specific phage by mouth. See also several WHO-sponsored studies in Pakistan in the 1970s, Monsur, K.A. et al., 'Effect of Massive Doses of Bacteriophage on Excretion of Vibrios, Duration of Diarrhoea and Output of Stools in Acute Cases of Cholera', *Bulletin of the World Health Organization* vol. 42, no. 5 (1970): 723–32.

Around the same time at the Pasteur Institute in Paris, oblivious of Twort and Hankin's work and for quite different reasons, a Canadian émigré Felix d'Hérelle discovered a 'microbe' that was 'antagonistic' to bacteria and that resulted in their death in liquid culture and death in discrete patches (he called them plaques) on the surface of an agar plate seeded with the bacteria (d'Hérelle, 1917, op. cit.).

The Tom Mix story and d'Hérelle and Eliava is wonderfully narrated in Kuchment, Anna, *The Forgotten Cure: The Past and Future of Phage Therapy*, Copernicus Books, Springer Science+Business Media, 2012.

During World War II, phage saved countless lives in countries that did not have access to antibiotics. The Soviet and Polish armies were using phage to treat the wounds of injured soldiers, and curing them from dysentery and typhus. The German military also used phage for wounds and diarrhoea, the evidence of which was found from medical kits seized from Rommel's North African forces. In the Soviet and East European theatre of war, there were sustained efforts in the use of phage therapy. See Summers, William C., 'The Strange History of Phage Therapy', *Bacteriophage* vol. 2, no. 2 (1 April 2012): 130–33.

Early results from experiments are showing promise for targeted and less toxic treatment of infections, especially for those who develop frequent infection from *E coli*. See Dalmasso, Marion et al., 'Three New *Escherichia Coli* Phages from the Human Gut Show Promising Potential for Phage Therapy', *PLoS one* vol. 11, no. 6 (9 June 2016).

For a review of the three decades (1929–40) of phage therapy developed and promoted by the Oswaldo Cruz Institute in Brazil, see Almeida, Gabriel Magno de Freitas and Lotta-Riina Sundberg, 'The Forgotten Tale of Brazilian Phage Therapy', *Lancet, Infectious Diseases* vol. 20, no. 5 (2020). Interestingly, the mass testing of phage products by the institute in 1924 preceded d'Hérelle's tests on cholera in India and the Soviet military's tests on troops. Brazil had also developed a better phage combination to fight staphylococcal infections but these were abandoned after the widespread adoption of antibiotics. For efficacy of phage to fight MRSA see Feng, Tingting et al. 'JD419, a *Staphylococcus aureus* Phage With a Unique Morphology and Broad Host Range.' *Frontiers in microbiology* vol. 12 602902. 22 Apr. 2021; and Berryhill, Brandon A et al. "Evaluating the potential efficacy and limitations of a phage for joint antibiotic and phage therapy of

Staphylococcus aureus infections.' *Proceedings of the National Academy of Sciences of the United States of America* vol. 118,10 (2021).

On the report of the 1932 Council on Pharmacy and Chemistry and the American Medical Association (1932) and the 1941 and 1945 reports published in the *Journal of the American Medical Association* which concluded that the use of bacteriophage for prevention and treatment therapy was confusing and often contradictory, thus marking the end for phage in US institutions and markets, see Kuchment, 2012, op. cit., and Summers, W.C. 1999, op. cit.

A popular medical textbook, considered a classic in its day, rather dismissively stated that all infectious diseases would be eradicated by the year 2000. See Cockburn, Aidan, *The Evolution and Eradication of Infectious Diseases*, Johns Hopkins University Press, December 1963.

In 1934, a British scientist, Robert Massey found the incidence and severity of bacterial blight disease in cotton, caused by a bacterium *Xanthomonas malvacearum*, was lower on lands that were frequently flooded by the Nile, and attributed this to phages that were transported by floodwaters. For a good overview, see Jones, J.B., A.M. Svircev and A.Ž. Obradović, 'Crop Use of Bacteriophages', in D.R. Harper, S.T. Abedon, B.H. Burrowes and M.L. McConville (eds), *Bacteriophages*, Springer, 2021.

On Twort's attempts to grow the smallpox virus using calf lymph and his discovery of the 'glassy transformation' of bacterial colonies in petri dishes, see Twort, F.W., 'An Investigation on the Nature of the Ultra-microscopic Viruses', *Lancet* (1915): 1241–43.

For the typed name on the death certificate, Thomas Edwin Mix, signed by Drs Seroggy, Smith and Dennis, dated 23 November 1931, 8 P.M., Diagnosis: Ruptured Gangrenous Appendix, Immediate Cause of Death: Peritonitis, see Johnston, Gary Paul, 'The Day Tom Mix Didn't Die!', *American Handgunner* (March/April 2018 Issue) Poway, California, USA.

The few large-scale publicly funded intervention studies were the WHO-sponsored studies in the former East Pakistan (now Bangladesh) in the 1970s. See Marcuk, L.M. et al., 'Clinical Studies of the Use of Bacteriophage in the Treatment of Cholera', *Bulletin of the World Health Organization* vol. 45, no. 1 (1971): 77–83.

The Council on Pharmacy and Chemistry, established in 1905 by the American Medical Association to set standards for drugs and lead the battle against quack remedies, undertook the evaluation of phage therapy in the mid-1930s. The voluminous report of the council concluded with an ambiguous assessment of the literature on phage therapy, acknowledging that there were both positive and negative results in the literature. There had been very good research that was conducted during the early 1940s by scientists like Rene Dubos at Harvard and physicians fighting typhoid in Los Angeles, but it largely disappeared. See Summers, 1993, op. cit.

In 2014 the WHO, realizing that the world is running out of effective ways to tackle common infections such as urinary tract infections and diarrhoea due to growing resistance, started to monitor AMR, and in 2016 initiated the Global Antimicrobial Resistance and Use Surveillance System (GLASS) to monitor the rise in AMR globally. Its most recent report, released in June 2020, lists eleven critical bacteria which are highly resistant to antibiotics. All except one (*Mycobacterium tuberculosis*, which causes TB) have an effective answer through the use of phage. The tuberculosis bacteria is difficult to treat because of the waxy outer layer which prevents phage from attacking it consistently. The report has dire warnings. The rate of resistance to ciprofloxacin, an antimicrobial frequently used to treat urinary tract infections, for example, globally varies from 8.4 per cent to 92.9 per cent across the thirty-three countries surveyed. It is increasingly becoming clear that some antibiotics are now well beyond their utility, and there are very few upon which we can depend.

Until the late 1980s, *S aureus* was susceptible to a wide range of antibiotics, until it acquired resistant genes, primarily with the help of viruses-smuggled genes from resistant neighbouring cells. See Haaber, Jakob et al., 'Bacterial Viruses Enable Their Host to Acquire Antibiotic Resistance Genes from Neighbouring Cells', *Nature Communications* vol. 7, no. 13333 (7 November 2016).

European tourists travelling to India, for example, who had absolutely no contact with the country's healthcare system still tested positive for carbapenemase-producing Enterobacteriaceae (CPE) after they came back from their trip. See Ruppé, Etienne et al., 'High Rate of Acquisition but Short Duration of Carriage of Multidrug-Resistant Enterobacteriaceae after Travel to the Tropics', *Clinical Infectious Diseases: An Official Publication of the Infectious Diseases Society of America* vol. 61, no. 4 (2015): 593–600.

A study published in June 2020 on the abundant and diverse variety of phages also found antibiotic-resistant genes (including CPE) were present in the river Ganga in India. This is indicative of an enormous public health problem looming over the whole of South Asia. See Ali, Saif et al., 'Influence of Multidrug Resistance Bacteria in River Ganges in the Stretch of Rishikesh to Haridwar', *Environmental Challenges* vol. 3 (April 2021). As on 15 June 2021, a search result on Google Scholar for Ganga 'antibiotic resistance', for a time period 2010 to 2021 yielded 1270 unique results, excluding citations.

The evolution of antibiotic resistance is a textbook case of what is often referred to as the Red Queen Effect (RQE). See Holmgren, Alicia M. et al., 'Outrunning the Red Queen: Bystander Activation as a Means of Outpacing Innate Immune Subversion by Intracellular Pathogens', *Cellular & Molecular Immunology* vol. 14, no. 1 (2017): 14–21.

Antibiotic resistance has largely been driven by two factors. See Huijbers, Patricia M.C. et al., 'A Conceptual Framework for the Environmental Surveillance of Antibiotics and Antibiotic Resistance', *Environment International* vol. 130 (2019).

Acinetobacter baumannii until the 1970s was perceived a low-grade pathogen. See Joly-Guillou, M.-L., 'Clinical Impact and Pathogenicity of Acinetobacter', *Clinical Microbiology and Infection: The Official Publication of the European Society of Clinical Microbiology and Infectious Diseases* vol. 11, no. 11 (2005): 868–73. But it has rapidly evolved to become a high-risk pathogen and is listed among the six top-priority microorganisms by IDSA. See Boucher, Helen W. et al., 'Bad Bugs, No Drugs: No ESKAPE! An Update from the Infectious Diseases Society of America', *Clinical Infectious Diseases: An Official Publication of the Infectious Diseases Society of America* vol. 48, no. 1 (2009): 1–12.

On the role of *Wolbachia* bacteriophage in reducing vector population, see Utarini, Adi et al., 'Efficacy of Wolbachia-Infected Mosquito Deployments for the Control of Dengue', *New England Journal of Medicine* vol. 384, no. 23 (2021): 2177–86.

On *Ralstonia* phage, see Kawasaki, Takeru et al., 'Genomic Characterization of *Ralstonia Solanacearum* Phage PhiRSB1, a T7-like Wide-host-range Phage', *Journal of Bacteriology* vol. 191, no. 1 (2009): 422–27.

Phages kill nearly half of all microbes in water. See Díaz-Muñoz, Samuel L. and Britt Koskella, 'Bacteria-phage Interactions in Natural Environments', *Advances in Applied Microbiology* vol. 89 (2014): 135–83.

IgM and IgNAR in sharks, for example, are functional analogues of mammalian IgG, and these cells are critical in immunization in fish (yes, farmed fish, like poultry and livestock receive shots!), and humans. See Flajnik, Martin F., 'A Cold-blooded View of Adaptive Immunity', *Nature Reviews Immunology* vol. 18, no. 7 (2018): 438–53.

A recent study on fish has found that phage can reduce bacterial virulence and therefore phage have wider application in disease management strategy in aquaculture. See Laanto, Elina et al., 'Phage-driven Loss of Virulence in a Fish Pathogenic Bacterium', *PLoS one* vol. 7, no. 12 (2012).

For microbes in petroleum, see Ollivier, Bernard and Magot, Michel (eds), *Petroleum Microbiology*, American Society for Microbiology (ASM), 2005; and Pannekens, Mark et al., 'Oil Reservoirs, an Exceptional Habitat for Microorganisms', *New Biotechnology* vol. 49, no. 25 (March 2019): 1–9.

Genital herpes or HSV2 perhaps was the first zoonotic (animal origin) disease of humans. See Bryson, Y. et al., 'Risk of Acquisition of Genital Herpes Simplex Virus Type 2 in Sex Partners of Persons with Genital Herpes: A Prospective Couple Study', *Journal of Infectious Diseases* vol. 167, no. 4 (1993): 942–46.

On the central role bacteriophages played in the development of molecular biology, see Cairns, J., G. Stent and J. Watson, *Phage and the Origins of Molecular Biology*, NY: Cold Spring Harbor Laboratory, 1966.

As of June 2021, the Cochrane registry of clinical trials, the US government website available at clinicaltrials.gov (15 July 2021) and WHO's ICTRP, have listed 142 completed and ongoing trials, which is an encouraging sign.

14 Quo Vadis?

On probable models of multidirectional evolution of viruses and exchange of genes across kingdoms and especially in fungi, bacterial plants and invertebrates, see Dolja, Valerian V. et al., 'Deep Roots and Splendid Boughs of the Global Plant Virome', *Annual Review of Phytopathology* vol. 58 (2020): 23–53.

8 per cent of the human genome is viral. See Horie, Masayuki et al., 'Endogenous Non-retroviral RNA Virus Elements in Mammalian Genomes', *Nature* vol. 463, no. 7277 (2010): 84–87; and Feschotte, Cédric, 'Virology: Bornavirus Enters the Genome', *Nature* vol. 463, no. 7277 (2010): 39–40. For more reviews of how bornaviruses invaded the mammalian cells, see Gifford, Robert J. 'Mapping the Evolution of Bornaviruses across Geological Timescales', *Proceedings of the National Academy of Sciences of the United States of America* vol. 118, no. 26 (2021). On reactivation of ancient viral genes inside the human genome, see Peddu, Vikas et al., 'Inherited Chromosomally Integrated Human Herpesvirus 6 Demonstrates Tissue-Specific RNA Expression *In Vivo* that Correlates with an Increased Antibody Immune Response', *Journal of Virology* vol. 94, no. 1 (12 December 2019).

When compared to our nearest ancestors, just 7 per cent of our DNA is unique to modern humans. See Schaefer, Nathan K. et al., 'An Ancestral Recombination Graph of Human, Neanderthal, and Denisovan Genomes', *Science Advances* vol. 7, no. 29 (16 July 2021).

On possible mutations and processes by a predecessor, SARS-CoV-2 from a mammal host acquired the ability to infect humans, see Liu, Kefang et al., 'Cross-species Recognition of SARS-CoV-2 to Bat ACE2', *Proceedings of the National Academy of Sciences of the United States of America* vol. 118, no. 1 (2021).

On comparison of genes of SARS-CoV-2, see Lehrer, Steven and Peter H. Rheinstein, 'Human Gene Sequences in SARS-CoV-2 and Other Viruses', *In Vivo (Athens, Greece)* vol. 34, no. 3 Suppl. (2020): 1633–36; and NCBI Genome website available at https://www.ncbi.nlm.nih.gov/genome/ (15 July 2021).

On the process of mutations occurring in SARS-CoV-2, see Callaway, Ewen, 'The Coronavirus Is Mutating—Does It Matter?', *Nature* vol. 585, no. 7824 (2020): 174–77.

On rate of infections, see Paul, J.H., 'Microbial Gene Transfer: An Ecological Perspective', *Journal of Molecular Microbiology and Biotechnology* vol. 1, no. 1 (1999): 45–50.

On how viruses influence microbial efficiency which regulates organic matter assimilation and carbon sequestration, see Bonetti, Giuditta et al., 'Implication of Viral Infections for Greenhouse Gas Dynamics in Freshwater Wetlands: Challenges and Perspectives', *Frontiers in Microbiology* vol. 10, no. 1962 (27 August 2019).

On the discovery of 1,95,728 marine viruses and entities, see Gregory, Ann C. et al., 'Marine DNA Viral Macro- and Microdiversity from Pole to Pole', *Cell* vol. 177, no. 5 (2019).

For a review of virus residents in the human body, see Liang, Guanxiang and Frederic D. Bushman, 'The Human Virome: Assembly, Composition and Host Interactions', *Nature Reviews Microbiology* vol. 19 (30 March 2021).

For a simple update on this complex science of immunotherapy, see https://www.cancerresearch.org/immunotherapy/treatment-types/oncolytic-virus-therapy (15 July 2021).

This Consensus Statement highlights the central role and global importance of microorganisms in climate change biology. See Cavicchioli, Ricardo et al., 'Scientists' Warning to Humanity: Microorganisms and Climate Change', *Nature Reviews Microbiology* vol. 17, no. 9 (2019): 569–86. Among the earliest papers that emphasised the role of microbes in climate mitigation was Danovaro, Roberto et al., 'Marine Viruses and Global Climate Change', *FEMS Microbiology Reviews* vol. 35, no. 6 (2011): 993–1034. For a paper which elegantly presents the cellular mechanisms of carbon sinking and oxygen production by phage-cyanobacteria interaction, see Campbell, Ian J. et al., '*Prochlorococcus* Phage Ferredoxin: Structural Characterization and Electron Transfer to Cyanobacterial Sulfite Reductases', *Journal of Biological Chemistry* vol. 295, no. 31 (2020): 10610–23.

For a comprehensive review of the benefits of virus and virus-based technologies, see Marintcheva, Boriana, *Harnessing the Power of Viruses*, Academic Press, 2017.

LIST OF ILLUSTRATIONS

Chapter 1: Bounty

p. 3: Nematode and Paramecium (Aria Lal), Micrasterias (Igor Siwanowicz/ Olympus Lifescience), *Aspergillus* with yeast (Aria Lal).

pp. 6–7: All virus images on this panel courtesy of Luke Jerram.

p. 8: Coronavirus images 1 and 2 (Unsplash, CDC), ultrastructural morphology of a coronavirus (Alissa Eckert, MSMI and Dan Higgins, MAMS, Luke Jerram).

Chapter 2: A Whole New World

pp. 16–17: Stelluti's Persio tradotto in verso sciolto e dichiarato from 1630 (Biodiversity Library); Grew's *The Anatomy of Vegetables*, 1672 (Internet archive); Hooke *Micrographia* (Royal Society); Malpighi's Opera Omnia and De formatione pulli (Complete Works), London 1686 (Internet archive); Redi's Experimenta circa generationem insectorum (Biodiversity Library); Ramesey's *Helminthologia* (Bodleian Library, Oxford), van Leeuwenhoek's heart muscle (Arcana, 1722, Wellcome Collection, Public Domain Mark); Joblot's hay infusion image from *Observations d'histoire naturelle*, Paris: Briasson, 1754–55 (Harvard Library).

p. 20: The Bolton & Watt factory (Louis Figuier [1867], Les Merveilles de la science ou description populaire des inventions modernes, 1819–94, p. 89), available at https://archive.org/details/FiguierMerveillesScienceBNF01_201409/ mode/2up. Drawing (Reichenbach's diary, courtesy Deutsches Museum, Munich).

p. 21: The 1826 one thaler coin with portraits of Reichenbach-von Fraunhofer (courtesy Museum Alte Münze).

p. 23: Man looking through the hand lens is enlarged from the frontispiece of Romeyn de Hooghe (1685) Titelprent voor 'Epistolae' [= 'Ontledingen en ontdekkingen van levende dierkens in de teel-deelen van verscheyde dieren, vogelen en visschen ... tot Londen in Engeland' of 'Anatomia seu Interiora rerum, ... ulteriore dilucidatione epistolis'?] of Antonie van Leeuwenhoek, available at archive.org.; illustration to show the use of the microscope in medicine, (Campani's microscopes), (Wellcome Collection). Attribution 4.0 International (CC BY 4.0); optics: a simple microscope concealed within a book binding, Engraving (Wellcome Collection). Attribution 4.0 International (CC BY 4.0).

p. 24: A travelling showman in the Netherlands showing animalcules to children, oil painting by a Dutch painter, ca. 1840 (Wellcome Collection). (CC BY-NC 4.0).

p. 25: A woman dropping her porcelain teacup in horror upon discovering the monstrous contents of a magnified drop of Thames water, revealing the impurity of London drinking water, coloured etching (W. Heath, 1828, Wellcome Collection). (CC BY-NC 4.0).

p. 28: Bas-relief on a monument located in Arbois, east France (courtesy Benoît Prieur - CC-By-SA).

p. 29: Card with notings and observations during preparation of the rabies vaccine by Louis Pasteur (courtesy Bibliothèque nationale de France).

p. 30: Two early images of pox by Ruska (B. von Borries, E. Ruska & H. Ruska [1938]: 'Bakterien und Viren in übermikroskopischer Aufnahme (mit einer Einführung in die Technik des Übermikroskops), in Klinische Wochenschrift', 17. Jg., 921–25; ectromelia virus (PHIL/CDC, 2012).

p. 31: Ruska's TMV (G.A. Kausche, E. Pfankuch & H. Ruska, 'Die Sichtbarmachung von pflanzlichem Virus im Übermikroskop', Naturwissenschaften vol. 27[1939]: 292–99); magnified image of TMV (ICTV).

Chapter 3: Supersize Me

p. 34: Timothy Rowbotham at work by himself, pink-stained Bradfordcoccus (courtesy Profs Raoult and Scolla), magnified mimivirus (IGS-CNRS UMR 7256).

p. 39: Comparison of RBC with other viruses (image by Aria Lal).

p. 40: Mimivirus and megavirus infected a single host, and mimivirus factory (IGS-CNRS UMR 7256).

Chapter 4: The Virus Is Us

p. 45: Image drawn for the book by Mihir Joglekar.

p. 48: Dissection of the uterus in the latter stages of pregnancy, showing the foetus, placenta and umbilical cord, coloured line engraving (W.H. Lizars, ca. 1827).

p. 51: Coelacanth (J. Rogers), foamy virus (ICTV).

p. 53: Papillomavirus (Luke Jerram).

p. 55: Koala (Deborah Tabart, Australian Koala Foundation).

Chapter 5: A Deep Control

p. 61: Hamelin Bay (Samantha Nichols), stromatolite (Pranay Lal).

p. 63: Image drawn for the book by Mihir Joglekar.

p. 64: *Prochlorococcus* and *Synechococcus* (Chisholm Lab, MIT), *Trichodesmium* and *Microcystis* (Pranay Lal), images of phages (ICTV).

p. 67: Global prevalence of oceanic micro-photosynthesisers (MIT Darwin Project, ECCO2, MITGCM/Oliver Jahn [MIT], Chris Hill [MIT], Mick Follows [MIT], Stephanie Dutkiewicz [MIT], Dimitris Menemenlis [JPL]).

p. 68: Algal bloom (Jacqueline Stefels), Phaeocystis (NIOZ).

pp. 70–71: First and third image (European Space Agency); second image (Earth Observatory, NASA, December 2015).

Chapter 6: Invaders, Hitch-hikers, Sentinels, Killers

p. 74: *Herpesviridae* and Anellovirus (ICTV).

p. 80: Ad36 (Luke Jerram).

pp. 82–83: Ape-human family (Elizabeth Daynes).

Chapter 7: A Spotty History of the Speckled Monster

p. 86: Smallpox (Luke Jerram), gerbil (IUCN).

p. 88: Pan-chen (Recherche sur les Superstitions en Chine [Research on Chinese Superstitions], Henri Dore, Shanghai, 1911–1920, from archive.org); textured Japanese watercolour (Wellcome Collection); Aztec mask (Pranay Lal); Sopona (Ashmolean Museum, Oxford).

p. 89: 1880 French engraving of children with three infections (Les Remèdes de la bonne femme, encyclopédie générale d'hygiène et de médecine usuelle).

p. 91: Śītalā mata (Wellcome Collection), Hārītī (Sridhar Padmanabhan, www. bronzesofindia.com), Jyestha (Wiki/CC), Mari-amma, Singapore (Pranay Lal).

p. 94: Gillray's caricature of a vaccination scene at St Pancras (Library of Congress, LC-DIG-ds-14062).

p. 95: Cruikshank caricature (Wellcome Library no. 11758i).

p. 96: Pandit Tudu and his wife (gouache drawing, Wellcome Collection, Attribution 4.0 International [CC BY 4.0], Ellis Original Leaf [Bibliothèque universitaire des langues et civilisations [BULAC]).

p. 97: Thomas Hickey, *Portrait of Three Princesses from Mysore*, 1805 (Sotheby's).

p. 101: Lithuanian mummy child (Kiril Cachovski); hand and leg of child from Prague (Petr Pajer); Viking grave (Västergötlands Museum); mummy of Ramesses V (G. Elliot Smith).

Chapter 8: Gut Feeling

p. 108: False colour composite of the tongue (Steven Wilbert and Gary Borisy, Forsyth Institute).

p. 111: *Prevotella, Bacteroides, Ruminococcus, Escherichia coli* (CDC/PHIL).

p. 115: crAssphage (ICTV).

p. 117: Bacteriophage-adherence-to-mucus, BAM (J.J. Barr et al., 'Bacteriophage Adhering to Mucus Provide a Non–host-Derived Immunity', *PNAS* vol. 110, no. 26 [25 June 2013]: 10771–76).

p. 119: Defence strategies of bacteria (Mihir Joglekar).

Chapter 9: A Virus Vanishes

p. 122: A. Hondius, *A Frost Fair on the Thames at Temple Stairs* (Museum of London).

p. 125: Danse Macabre (La grande danse Macabre des hommes et des femmes, 1862); a Dutch doctor looking through a flask of cloudy urine (image of sweating sickness from Dutch republic is from the book by Salomon van Rusting, *Het Schouw-Toneel des Doods; waar op na't leeven vertoont word De Dood op den Throon des Aard-Bodems*, or The Scene of Death; where after life shows death on the throne of the bottom of the earth becomes reigning over all states and nations or the [Danse Macabre]: Heerschende over alle Staaten en Volkeren, Amsterdam: Johannes Rotterdam, 1707).

p. 126: Mozart's death certificate (Domarchiv St Stephan, Vienna [AT], registered archive of the cathedral parish, Bahrleihbuch [book of funerals] 1791, folio 337, 6 December 1791).

p. 129: Bank vole (Claire Spelling/Flickr), hantavirus (CDC).

Chapter 10: Beauty

pp. 134–35: Wild tulips, Kazakhstan (Russell Scott).

pp. 136–37: Paintings of Brueghel (Museo Del Prado); van den Berghe (Christie's), de Gheyn II (rkd.nl), de Champaigne (Tessé Museum), van Spaendonck (Sotheby's).

p. 139: Brueghel the Younger satire (Alamy).

pp. 140–41: Ottoman fabric with tulips, silk and silver lamella from around the 1650s (Topkapi Museum), glass vessel (Cairo Museum) and Turkish miniature (Topkapi Museum).

p. 142: Bosschaert, still life of four tulips in a Wan-Li porcelain vase (WikiCommons).

p. 143: Potyvirus (ICTV).

Chapter 11: How a Virus Saved a Giant

p. 146: Chestnut tree (Dumbarton Oaks Museum Library and Collection), dead trees (USDA), map of chestnut blight redrawn by Aria Lal from G. Ashley, USDA, 1962 and http://www.fao.org/3/x5348e/x5348e04.jpg.

p. 153: *Cryphonectria* on tree and petri dish (John Plischke), CHV-1 (ICTV).

Chapter 12: Zombies

p. 157: Toxoplasma (Jun Liu/Olympus).

p. 160: Kuru-affected boy and father (UK MRC Prion Unit and Papua New Guinea Institute of Medical Research).

p. 161: Normal brain cross-section, kuru-infected brain, CJD or mad cow disease in humans, and MCD-infected brain tissue of a cow (Lane Medical Library, Stanford University, CC BY-NC 4.0).

p. 164: Cranach woodcut (Wiki Commons, CC0 1.0).

p. 166: Emerald wasp attacking a cockroach (Aria Lal), young wasp (Alamy) and polydnavirus (ICTV).

p. 168: Baculovirus (ICTV), caterpillar oozing (Aria Lal), infected caterpillar (Kathy Keatley Garvey), emerald wasp (Alamy).

p. 170: Yellowstone (Pranay Lal), fungus and rosette grass (Wiki Commons), CThTV (ICTV).

Chapter 13: Enemy's Enemy

p. 176: *Vibrio* (Pranay Lal), copepod (Rogelio Moreno Gill/Olympus), CTXΦ and Vp5virus phage (ICTV).

pp. 181, 184: Sgt Pepper's Lonely Hearts Club Band and Elizabeth Taylor (Getty images).

p. 187: Phage application on infected foot (Ramaz Katsarava, Georgian National Academy of Sciences Tbilisi, Georgia).

p. 188: *Wolbachia* inside ovary of mosquito (World Mosquito Program), WO bacteriophage (ICTV).

p. 189: Different ways in which *Wolbachia* infects male and female mosquitoes (Aria Lal).

p. 192: *Ralstonia solanacearum* and phage (Belén Álvarez, María M. López and Elena G. Biosca, 'Biocontrol of the Major Plant Pathogen *Ralstonia solanacearum* in Irrigation Water and Host Plants by Novel Waterborne Lytic Bacteriophages', Microbiology, 6 December 2019, https://doi.org/10.3389/fmicb.2019.02813).

p. 193: Tomato and papaya farm and wilting banana leaf from Colombia (Ministerio de Agricultura y Desarrollo Rural).

INDEX

Bouquet, Henry, 89
Bradford, 33
    pneumonia outbreak, 33, 34pc
Bradfordcoccus, 34–35
brain and nerves, degenerative illness
    of, 159
brain, development of, 50
brain-meddling pathogens, 158
Brando, Marlon, 180
Brazil, 81, 88pc, 103, 183, 189pc,
    192pc, 221, 239
Brazilians, 158
breast milk, 49
British, 89
British Isles, 122
British Medical Journal, 129
British vineyards, 123
Broad Institute at MIT, 110
Bronx Zoo, New York City, 148
Brueghel, Jan, 139pc
Brussels, 129, 179, 183, 228
bubonic plague, 179
buffalopox, 103, 221
building blocks of life, 66
'bunching' in flowers, 144
Bunyavirales, 128
butterflies, 149, 168, 188pc

Cabernet Sauvignon grapes, 144
Cafeteria roenbergensis, 36
Calais (France), 65
calcium, 60, 64, 214
calcium carbonate, 64
calcivirus, 109
Calcutta, 179
calf's lymph, 177
Cambridge University, 124
camellias, 144
camelpox virus, 86, 103
cancers
    breast cancer, 51
    cells, 56
    cervical, 52, 53pc
    colorectal, 51
    control, 199

    liver, 51
    melanomas, 51
    ovarian, 51
    pancreatic, 51
    prostate, 51
cankers, 150–52, 234
cannibalism, 159, 169, 171
capillaries, 19
capsid, 63pc
carbon, 62, 64, 66, 70, 71, 214
carbon capture, 149
carbon dioxide, 62, 66, 70
carbon-fixing, 67pc
carbon-oxygen cycle, 197
carbon sinks, 62
    in deep seas, 71pc
Caribbean islands, 162
Carillo, Leo, 180
Carrara (Italy), 65
Carroll, Lewis, 190
catatonia, 159
caterpillar cadavers, 168
cathode ray tube (cathode gun), 29
cats, 54, 211
cattle, 28, 98, 159, 162, 174
Caudovirales, 114
Cayley, Dorothy, 138
cell membranes, 119pc
cellular life, 5
Centers for Disease Control and
    Prevention (CDC), US, 8pc, 156,
    188
Central Africa, 77, 86pc
central nervous system, 36
cerulean warbler, 149, 233
cervical cancer, 52, 53pc
cervid endogenous retrovirus (CrERV),
    54
Ceylon, 96pc
chaalisas (or forty hymns) Hindu texts,
    104
chalk, 65
Chamberland, Charles, 28
Chamberland filter candles, 26, 28
Chaos, 7

French Academy of Sciences, 29pc
*Frost Fair on the Thames at Temple Stairs,
A* (1684), 122pc
frost fairs (London), 123
Frost, Robert, 148
Fuchs, Shai, 106–7
full break, 138
functionally extinct, 149, 234
fungi, 3pc, 185, 202, 231, 247
    Arthrobotrys, 3
    Aspergillus, 112, 222
    diseases caused by
        Dutch elm disease, 147
        white pine blister rust, 147
    human-associated, 112, 222
    spores, 3pc, 15, 17pc, 147, 169,
        202
    thread-like, 4

Gabon, 41, 207
Galway, Ireland, 124
game hunting, 149
Gandhi, Mahatma, 180
    satyagraha, 179
Ganga River, 176
gargantuan geological processes, 59
gastroenterology, 109
gastrointestinal diseases, 192
genes
    2 per cent hypothesis, 100
    adding, editing and deleting, 196
    encoding the globin, 50
    language of, 99
genetically modified mice, 161
genetic code, errors in replicating, 195
genetic diversity, of viruses, 114
genetic Frankensteins, 196
genetic mutations, 195
gene transfers, episodes of, 55
genital warts, 52, 53pc
Genoa, Italy, 150–51
genomes
    mapping of, 10
    sequencing of, 11
genome-sequencing technologies, 114

rapid genetic sequencing
    techniques, 191
genus (*Castanea*), 145
geological processes, 59–60, 197, 215
George Eliava Institute of
    Bacteriophages, Microbiology and
    Virology, 183, 185
Georgia, 82pc, 103, 145, 150, 152,
    183, 191
gerbil, 85–86
    immune system, 86
    naked-soled, 218
    northern savannah gerbil
        (*Gerbilliscus kempi*), 86
    tatera poxvirus, 86
German for 'severe military fever', 126,
    126pc
German physicists, 29
    Berlin Group, 29
Germany, 21pc
gestation period, 50
Gettysburg address, 87
Ghebreyesus, Tedros Adhanom, 104
giant kelp, 66
giant viruses, 34pc, 36–38
gibbon ape leukaemia virus (GALV),
    53
gibbons, 53, 211
Giger, H.R., 166
Gillray, James, 94pc
gladioli, 144
glass
    mirrors, 14
    optical instruments, 14, 21pc, 22
    telescope, 14, 204
glassmaking, in Bavaria, 21pc
Glendale Airport, 182
Global Commission for the
    Certification of Smallpox
    Eradication, 104
global warming, 70
Gloucestershire, 92–93
glucose (blood glucose), 80
glycines, 161
glycoproteins, 117pc